Aluminum Surfaces

ZAHNER'S ARCHITECTURAL METALS SERIES

Zahner's Architectural Metals Series offers in-depth coverage of metals used in architecture and art today. Metals in architecture are selected for their durability, strength, and resistance to weather. The metals covered in this series are used extensively in the built environments that make up our world and are also finding appeal and fascination to the artist. These heavily illustrated guides offer comprehensive coverage of how each metal is used in creating surfaces for building exteriors, interiors, and art sculpture. This series provides architects, metal fabricators and developers, design professionals, and students in architecture and design programs with a logical framework for the selection and use of metallic building materials. Forthcoming books in *Zahner's Architectural Metals Series* will include Stainless Steel; Aluminum; Copper, Brass, and Bronze; Steel; and Zinc surfaces.

Titles in *Zahner's Architectural Metals Series* include:

Stainless Steel Surfaces: A Guide to Alloys, Finishes, Fabrication and Maintenance in Architecture and Art

Aluminum Surfaces: A Guide to Alloys, Finishes, Fabrication and Maintenance in Architecture and Art

Aluminum Surfaces

A Guide to Alloys, Finishes, Fabrication, and Maintenance in Architecture and Art

L. William Zahner

WILEY

Cover image: © L. William Zahner
Cover design: Wiley

This book is printed on acid-free paper.

Copyright © 2020 by John Wiley & Sons, Inc. All rights reserved.

Published by John Wiley & Sons, Inc., Hoboken, New Jersey.

Published simultaneously in Canada.

No part of this publication may be reproduced, stored in a retrieval system, or transmitted in any form or by any means, electronic, mechanical, photocopying, recording, scanning, or otherwise, except as permitted under Section 107 or 108 of the 1976 United States Copyright Act, without either the prior written permission of the Publisher, or authorization through payment of the appropriate per-copy fee to the Copyright Clearance Center, 222 Rosewood Drive, Danvers, MA 01923, (978) 750-8400, fax (978) 646-8600, or on the web at www.copyright.com. Requests to the Publisher for permission should be addressed to the Permissions Department, John Wiley & Sons, Inc., 111 River Street, Hoboken, NJ 07030, (201) 748-6011, fax (201) 748-6008, or online at http://www.wiley.com/go/permissions.

Limit of Liability/Disclaimer of Warranty: While the publisher and author have used their best efforts in preparing this book, they make no representations or warranties with respect to the accuracy or completeness of the contents of this book and specifically disclaim any implied warranties of merchantability or fitness for a particular purpose. No warranty may be created or extended by sales representatives or written sales materials. The advice and strategies contained herein may not be suitable for your situation. You should consult with a professional where appropriate. Neither the publisher nor the author shall be liable for damages arising here from.

For general information about our other products and services, please contact our Customer Care Department within the United States at (800) 762-2974, outside the United States at (317) 572-3993 or fax (317) 572-4002.

Wiley publishes in a variety of print and electronic formats and by print-on-demand. Some material included with standard print versions of this book may not be included in e-books or in print-on-demand. If this book refers to media such as a CD or DVD that is not included in the version you purchased, you may download this material at http://booksupport.wiley.com. For more information about Wiley products, visit www.wiley.com.

Library of Congress Cataloging-in-Publication Data:

Names: Zahner, L. William, author.
Title: Aluminum surfaces : a guide to alloys, finishes, fabrication and
   maintenance in architecture and art / L. William Zahner.
Description: First edition. | Hoboken, New Jersey : John Wiley & Sons, Inc.
   [2020] | Series: Zahner's architectural metals series | Includes
   bibliographical references and index. |
Identifiers: LCCN 2019016600 (print) | LCCN 2019017323 (ebook) | ISBN
   9781119541776 (Adobe PDF) | ISBN 9781119541752 (ePub) | ISBN 9781119541769
   (pbk. : acid-free paper)
Subjects: LCSH: Aluminum—Finishing. | Aluminum—Surfaces. | Aluminum
   coatings. | Architectural metal-work. | Art metal-work.
Classification: LCC TS555 (ebook) | LCC TS555 .Z28 2020 (print) | DDC
   673/.722—dc23
LC record available at https://lccn.loc.gov/2019016600

Printed in the United States of America

V10013217_082019

To a great artist and good friend, Jan Hendrix.

Contents

	Preface	xi
CHAPTER 1	Introduction to Aluminum (Aluminium)	1
	Introduction	1
	Aluminum as a Design Material	7
	Environmental and Hygienic	9
	History of the Metal	14
	The Origins of Modern Production	16
	Aluminum in Art	22
	Production Process Today	25
	Aluminum versus Aluminium	29
	Comparisons Between Architectural Metals	30
CHAPTER 2	Aluminum Alloys	33
	Choosing the Correct Alloy	35
	The Initial Mill Casting	35
	Commercial Pure Aluminum Alloy A91xxx	38
	Alloy Designation System	39
CHAPTER 3	Surface Finishing	95
	The Mill Finish Surface	96
	Directional Finishes	102
	Non-Directional Finishes	102
	Mirror Finish	104
	Bright Dipping and Electropolishing	109
	Electropolishing Aluminum	111
	Anodizing	111
	History of Anodizing	115
	Anodizing Process	116
	The Anodized Surface Layer	119
	Variations in the Anodized Coating	126
	Anodizing Quality	126
	Defects That Can Appear After Anodizing	127
	Effects of Lubricants	127
	Structural Streaks	128

	Managing Expectations of the Anodized Surface	128
	Coloring by Means of Anodizing	130
	Design Considerations	144
CHAPTER 4	The Aluminum Surface Finish: Meeting Expectations	161
	The Natural Finish	162
	The Anodic Finish	167
	The Point of Range Samples	171
	Color	173
	Flatness	178
	Cleaning the Surface Over Time	182
CHAPTER 5	Designing with the Available Forms of Aluminum	187
	Basic Forms of Aluminum	189
	It Begins at the Aluminum Mill: The Heat	190
	Sheet and Plate	190
	Supply Constraints	191
	Plate	193
	Coil and Sheet	197
	Foil	200
	Clad Aluminum: Alclad	202
	Extrusion	203
	Extrusion Process	207
	Designing the Shape	208
	Cross-Section Design Criteria	208
	Extrusion Dies	211
	Aluminum Pipe and Tube	213
	Aluminum Rod and Bar	215
	Aluminum Structural Shapes: Angles, Channels, Tees, and I-Beam Shapes	215
	Aluminum Wire	219
	Aluminum Wire Mesh	219
	Aluminum Expanded Metal	220
	Perforated Aluminum and Embossed Aluminum	222
	Cast Forms	224
	Sand Casting	227
	Permanent Mold Casting	228
	Die Casting	228
	Aluminum Foam	229
CHAPTER 6	Fabrication	231
	Challenges with Aluminum Fabrication	233
	Handling and Storage	235
	Forming	254

		Fastening	269
		Soldering and Brazing	271
		Welding	272
		Casting	279
		Rapid Prototype	283
	CHAPTER 7	Corrosion Characteristics	285
		Natural Weathering: Influences on Performance	287
		Marine Environment	289
		Urban Environment in the South	290
		Urban Environment in the North	291
		Rural Environment	292
		Corrosion of Aluminum Surfaces	293
		Water Staining	310
		Fingerprinting	311
		Expectations of Various Environmental Exposures	311
		Cleaning	313
		Protective Measures	314
		Paint Coatings	315
		Storage and Handling	322
	CHAPTER 8	Coping with the Unexpected	325
		Introduction	325
		The Principal of Life Cycle	326
		Mill Constraints	327
		Potential Issues with Sheet and Plate	329
		Dents and Scratches in Anodized and Painted Aluminum	332
		Graffiti	333
		Flatness	334
		Stiffener Show-through	339
		Issues Encountered with Casting Aluminum	340
		Issues Encountered with Aluminum Extrusions	341
		Crazing of the Surface	344
		Causes of Tonal Differences	347
		Range Samples	348
		Welding of Anodized Assemblies	350
		Spalling	351
		Water Stains	351
		Inclusions Visible After Polishing Plate	352
	APPENDIX A	Valuable Information and Specifications for Aluminum in Art and Architecture	353

APPENDIX B	Tempers on Wrought Sheet, Extrusion, and Plate for Alloys Considered for Use in Art and Architecture	355
APPENDIX C	European Specifications Relevant to Art and Architecture	357
APPENDIX D	Alloy Designations for Wrought Aluminum Alloys Used in Art and Architecture	361
	Further Reading	363
	Index	365

Preface

Several years ago, I was a speaker at a symposium put on by the American Metals Market in Pittsburgh. My talk described the use of metals in art and architecture. Later that evening, I had the opportunity to sit next to an executive of a large aluminum company during dinner. Over drinks, we got into a discussion of the metal, aluminum, and how it is used in architecture. We talked about some of the amazing work that had been accomplished, much in the early years of the metal when designers embraced aluminum as a decorative skin for buildings. Imagine how building construction changed from an emphasis on large, heavy limestone, brick, or granite façades to the modern look of lightweight, waterproof, shiny surfaces offered by this new modern material called aluminum.

I asked him why, of late, it is often difficult to get good, consistent quality aluminum plate for anodizing on architectural projects. He reached forward with his finger, traced the boundary of the placemat under his glass of wine, and said, "Let's suppose this represents the entire aluminum industry in America." Then he traced out an area about the size of a postage stamp in one corner and said, "This is the architectural market; the rest is cans." He went on to say that the architectural market is important, but it is more difficult and presents more risk—not so much on the part of the producer of the initial casting, but for companies that distribute the metal for architectural projects. In these cases the risk is very high. Cans are less risky, and the market is very large.

What was not said by this industry leader was that most of the aluminum used in the United States today is generated from secondary metal, that is, recycled aluminum has outpaced the use of primary aluminum produced from refined ore. This secondary metal is outstanding from an environmental standpoint. Far less energy is consumed. There is less concern regarding environmental controls on greenhouse gases or hazardous fluoride emissions. The necessary mechanical properties and corrosion resistance are still achieved with the alloys generated from heavier recycled content. The aluminum can industry, automotive industry, and transportation industry do not have as great of concern with the trace amounts of impurities that can be included in the alloys.

However, the demands of the architectural world can be significant. Small traces of impurities, such as iron, can have a subtle effect on the end color tone of the aluminum, which can be enhanced when the metal is anodized. Controlling these trace impurities drives the costs higher in production and subsequent processing of architectural aluminum. More often the aesthetic issues do not manifest in the metal's surface until after all the costs of final production are already spent. This can be devastating after all the effort put into producing a magnificent project.

With new techniques and a design sensibility, we can find ways of working with the potential of tonal differences in aluminum surfaces. We can make these concerns a thing of the past.

It is knowledge and experience that is needed. Rather than relaxing the architectural design criteria, work can be done with the aluminum to understand the nuances of the natural material.

More often than not, aluminum in art and architecture is coated with high-quality coats of paint. Aluminum makes an outstanding substrate to receive paint. The paint actually will perform better because of the inert nature of aluminum. But this book is not about paint coatings but about the metal itself. Aluminum's intrinsic beauty is combined with its incredible corrosion resistance and ductile yet strong performance.

Like stainless steel, aluminum is one of those surface materials expected to not change drastically from weathering as the ravages of time and the environment interact with the metal's surface. This aspect of architectural metals, an unchanging appearance with time and exposure, is the expectation of two metals commonly used in art and architecture: stainless steel and aluminum. Gold would be another desirable metal, but, short of gold leaf, few architectural or art projects have the budgets for gold-leaf buildings. Titanium is making inroads into architecture, and as its price comes down it may be a more formidable competitor for use in art and architecture. But the other common metals in use—steel, weathering steel, copper and copper alloys, zinc—all will change with time and exposure. Certainly aluminum changes, but at a very slow rate in most of our built environments.

Aluminum is an amazing metal. It is as if it were created with the express purpose of furnishing us with a material to move mankind into a new and bold future.

CHAPTER 1

Introduction to Aluminum (Aluminium)

Al – Aluminum

... seems to have been created with the express purpose of furnishing us with the material for our projectile.

—Jules Verne, *From the Earth to the Moon*

INTRODUCTION

Aluminum is the metal of design.

Widely used in architecture, furniture, transportation, and increasingly more often by artists as a sculptural material, aluminum in its various alloying forms offers the designer a vast and diverse assortment of choices. Aluminum is relatively new as a material for design but the forms it can take are incomparable.

Aluminum possesses its own nuances that give it distinctively different characteristics from those of other metals. These subtleties offer the designer and artist unique ways of working with the metal and creating intriguing and long-lasting forms and colorful surfaces that will stand the test of time in the environment we live in.

Aluminum is the most abundant metallic element on the surface of the earth, making up nearly 8% of the minerals found on the earth's surface. Unlike copper, silver, gold, and platinum—metals that can be found in their pure elemental form—aluminum, due to its atomic makeup, is only found in combinations with other substances. When the earth was formed, aluminum was a light metal and was quick to bond with other elements, in particular oxygen and silicon, the two most abundant elements on the earth's surface. The low density of aluminum allowed it to float on the surface of denser metals, such as iron and nickel. As the earth's surface cooled and formed, aluminum combined with other metals and nonmetals to make up a sizeable portion of the earth's surface.

Most abundant elements on the earth's surface	
Oxygen	47%
Silicon	28%
Aluminum	8%
Iron	5%
Calcium	4%

Aluminum is a common element in many minerals. Because it exists in combinations of oxides and silicates along with other metals, it would be difficult to find areas where there is no aluminum in the rocks and minerals that make up this planet. Many of the most precious gemstones contain aluminum. Topaz, ruby, tourmaline, emerald, turquoise, lapis, garnet, and even forms of jade contain aluminum in their chemical makeup (Figure 1.1).

FIGURE 1.1 Image of gemstones containing aluminum compounds.

Paradoxically, this light, soft metal when combined with only oxygen can become one of the hardest substances known to man. Corundum is simply aluminum oxide, Al_2O_3, yet this mineral is nearly as hard as diamond. Dark corundum, known as emery, is an abrasive used in grinding and polishing other minerals and metal surfaces. However, if impurities of other metals are included in corundum, then the mineral takes on color. Rubies get their red color from trace amounts of chromium and iron in the corundum mineral. Beautiful blue sapphire is corundum with trace amounts of titanium and iron.

Aluminum, in one mineral form or another, has been used for centuries in one of the constituents that make up clay. At one time it was referred to as "the metal of clay." Clay consists of aluminum in combination with silicon and oxygen, as well as with other metals. These aluminosilicates, along with other substances, were the basis for pottery more than 7000 years ago. Today, the bricks we see in many of our buildings are actually composed of aluminum silicates.

Bauxite is the most widely mined ore of aluminum. Bauxite is a mixture of several aluminum minerals, along with other clay minerals. The aluminum minerals that make up a large portion of bauxite are gibbsite, boehmite, and diaspore. Each of these minerals is an oxyhydroxide of aluminum.

Gibbsite	$Al(OH)_3$
Boehmite	$AlO(OH)$
Diaspore	$AlO(OH)$

Africa, Indonesia, and Australia are some sources of the ore known as bauxite, which is found on the surface of the earth in tropical regions around the world. The severity of the weathering processes over centuries in these tropical regions had the effect of leaching the silicates from the aluminosilicate rocks and leaving hydrated aluminum oxides behind. Bauxite is the original source of all aluminum today.

The abundance of aluminum on the earth did not make it easily available to mankind. For most of human existence, aluminum was locked away in clay and rocks. The atomic makeup of this element make it difficult to free the metal from the surrounding material (Figure 1.2).

Aluminum will readily shed the three electrons that reside in its outer shell. Aluminum sheds these three electrons in order to arrive at a balanced outer shell. Metals usually have one, two, or three electrons in their outer shell, which gives them a positive valence corresponding to this electron count. Aluminum has a valence of 3^+, which makes it quick to bond with other substances, such as oxygen, which has a valence of 2^-. A common aluminum compound found in nature is its oxide Al_2O_3, which is two atoms of aluminum with a valence of 6^+ and three oxygen atoms with a valence of 6^-. All matter on earth seeks out a balanced charge.

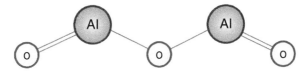

Aluminum Oxide Al_2O_3

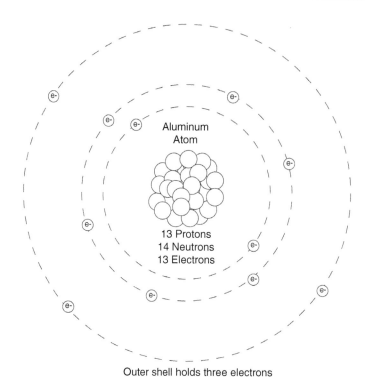

FIGURE 1.2 Aluminum atom.

The three electrons are given up to oxygen atoms in this case, because aluminum has a very low ionization energy. In other words, very little energy is needed for aluminum to shed its outer three electrons. It wants to share them and usually oxygen is readily available to accept. This low ionization energy causes aluminum to rapidly form an oxide when exposed to air. When this bond with oxygen is formed it is very stable and difficult to break. This oxide, which forms on the surface of aluminum, is clear, inert, and hard. The layer resists change when confronted with corrosive substances in the normal environment. This oxide is what gives aluminum its excellent corrosion resistance.

Aluminum reactions nearly always involve the loss of three electrons, but never more than three, as removal of a fourth electron would require significant energy. It is the strong triple electron bond that aluminum makes with other substances that makes it difficult to isolate the element. Effective energy sources were finally employed to release aluminum from its bonds with other elements.

Aluminum had very limited use until the early 1900s. The difficulties in refinement caused the metal to be no more than an expensive novelty. Back in the mid-1800s, aluminum was considered a precious metal, comparable in price with gold or platinum. Until the beginning of the twentieth century, only small amounts of aluminum had ever been isolated to the point where objects could be created from it.

Aluminum is lightweight, with a low density, 35% that of iron and around 30% of copper. It is not the lightest industrial metal—magnesium is 33% lighter than aluminum, but is rarely used in art and architecture. Aluminum is an excellent conductor of heat and electricity. Aluminum's conductivity is approximately 63% that of copper, because of the three electrons in the outer shell. They enable the transfer of electrical current and thermal energy through the metal, but because the atomic number is low—19 for aluminum compared to 29 for copper—the atomic radius of aluminum is less. This means the electrons for aluminum are slightly more bound to the aluminum nucleus and less free to move around and aid in the conductivity of electricity than the electrons in the larger copper atom. Most metals have free electrons in their outer shells that are free to move about in the crystal structure of metal atoms. Silver, copper, and gold have higher conductivity than aluminum, but the cost and density of aluminum make it an optimal conductor of electrical current.

Aluminum is nonmagnetic and in its pure form has a silvery-white metallic color because it reflects a large portion of the visible spectrum. Aluminum has a low melting point, 463–670°C. The melting point is half that of copper and a third of the melting point of steel, making aluminum a low-energy recyclable metal and a low-temperature casting metal (see Table 1.1).

TABLE 1.1 Melting points of various metals.

Metal	Liquidus (°C)
Aluminum	658
Brass—yellow	930
Copper	1084
Iron—wrought	1540
Steel	1483
Stainless steel	1510
Tin	232
Titanium	1670
Zinc	419

Aluminum alloys melt at a range of temperatures. Melting begins at a lower temperature then achieves full liquidus at the high end of the range.

Aluminum falls between magnesium (atomic number 12) and silicon (atomic number 14), which are two elements that mix well with aluminum and are common to many alloys that make up the aluminum family. All are found in abundance on the earth's surface, in combinations with other elements that form the rocks and minerals that make up this planet.

Aluminum **Element 13**
Atomic number 13
Crystal structure Face-centered cube
Main mineral source Bauxite
Color Silver-white
Oxide Clear and nonstaining
Density 2700 kg m^{-3}
Specific gravity 2.7
Melting point 658°C
Thermal conductivity 210 W/m °C
Coefficient of linear expansion 24×10^{-6}/°C
Electrical conductivity 62% International Anealed Copper Standard (copper = 100%)
Modulus of elasticity 70 GPa

Aluminum is the most abundant metal on the Earth and makes up 8% by weight of the earth's surface.

High strength-to-weight ratio
Excellent ductility, deep forming ability
High fracture toughness
High elasticity/resiliency under shock loading
Soft edge
Nontoxic
Superior corrosion resistance in many natural environments.

Finishes
Mill specular and nonspecular
Polished satin and mirror
Glass bead
Anodized
Paint coatings

Artificial patina Dark gray to black
Bright appearance: Aluminum reflects over 90% of the visible spectrum
Reflectance of ultraviolet Excellent
Reflectance of infrared Excellent
Relative cost Low
Strengthening Heat strengthening and cold working
Recycle ability Very easily recycled. A secondary industry is devoted to the recycling of the metal. Only 5% of the energy that is used to make new aluminum is needed.

Welding and joining	Aluminum can be welded and brazed. Strength is affected by heat of weld. Fasteners are usually stainless steel or aluminum.
Casting	Aluminum can be sand cast, permanent mold cast, and die cast.
Plating	Aluminum can be electroplated, but with difficulty. Requires specialized techniques.
Etching and chemical milling	Aluminum can be etched and chemically milled.
Unique behavior	Aluminum will increase in strength as temperature decreases.

Aluminum possesses properties that enable it to take numerous forms and shapes far more readily and with less energy consumption than other metals. Aluminum can be extruded through dies and shaped through casting techniques; sheets, plates, and foils are all forms that aluminum can take.

Aluminum foils can be rolled as thin as 0.04 mm (0.0017 in.), less than the thickness of a human hair. Aluminum wire can be drawn out to as little as 0.1 mm (0.004 in.). On certain alloys, heat treatment processes can increase the strength of aluminum to 662 MPa (96 ksi), levels that exceed the strength of common steels.

Today aluminum is one of the most versatile metals known and one of the most commonly available for use by artists and architects. In the past century, no other metal has had such an expansive effect on architecture, industry, and transportation. Couple aluminum's abundance with its excellent corrosion resistant properties and you have a metal designed for mankind's needs.

Aluminum possesses its own nuances that give it distinctively different characteristics from those of other metals. These subtleties offer the designer and artist unique ways of working with the metal and creating intriguing and long-lasting forms and colorful surfaces that will stand the test of time in the environment we live in.

Aluminum possesses properties that enable it to take numerous forms and shapes far more readily and with less energy consumption than other metals.

ALUMINUM AS A DESIGN MATERIAL

As a metal to be shaped and finished for a great number of uses, aluminum is like no other design material in existence. However, aluminum in its highly pure state is very weak. The tensile strength of pure aluminum is approximately 90 MPa (13 ksi), not much stronger than some plastics. However, cold-working the metal can double its strength. Adding alloying elements as well as cold-working will increase the strength even further. Heat treatment processes and aging will take aluminum to the strengths of steels.

Metals are crystalline materials that exist in the solid state, and aluminum is no exception. In order to plastically deform into shapes, the crystals must have the ability to slip or dislocate as stress is applied. In wrought forms, aluminum crystals are in layers and these layers dislocate under stress

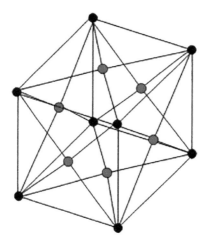

FIGURE 1.3 Face-centered cubic structure of aluminum.

and allow the metal to plastically deform. When this slipping is interrupted, for example, by introducing another alloying element into the crystal matrix, the metal is strengthened (Figure 1.3).

Controlling the grain size, which can be done by heating the metal and then cooling it at a particular rate, reduces the continuity of its crystal layers. The metal is strengthened through a process called strain hardening, which creates new grains and mixes them up. Cold working puts aluminum through plastic deformation, which increases the hardness and strength of the metal by transferring energy into the metal. The energy is stored in the strained metal crystals in the form of deformations of the crystals. Heat treatments can remove some of this strain hardening and reduce the strength of the metal. When aluminum is held at a certain temperature for a period of time, a process called atomic diffusion occurs in which the atoms move around. This relieves some of the built-up energy of the internally strained crystals as the atoms move from regions of high strain to regions of lower strain.

Aluminum will lose strength at elevated temperatures, but strength will increase at temperatures below freezing without loss of ductility. Other metals may become brittle at extremely low temperature, but aluminum actually performs well and even gains strength.

At high temperatures, such as during a fire, aluminum and its alloys can actually provide levels of protection.[1] A mass of aluminum takes twice as much energy as a similar mass of steel to raise its temperature a single degree. This is due to the high specific heat capacity of aluminum. The thermal conductivity of aluminum is significantly higher than steel and will conduct heat away. The reflectivity of aluminum in the infrared wavelength climbs above 90% and will reflect away the heat energy more effectively than many other materials.

Aluminum, rolled thin into sheets and foils, is an excellent reflector of radiant energy over the entire range of wavelengths from ultraviolet to infrared. This property is used, for example, to reflect heat back to foods that are wrapped in thin aluminum foils.

---

[1] Kaufman, J.G. (2016). *Fire Resistance of Aluminum and Aluminum Alloys and Measuring the Effects of Fire Exposure on Properties of Aluminum and Aluminum Alloys*. ASM International.

It will also reflect the electromagnetic waves of radio and radar. Thin aluminum foil or even aluminum mesh made from woven aluminum wire can reflect radio waves. Because of aluminum's ability to conduct electricity, it will block radio wavelengths as a Faraday cage effect is developed. Thin aluminum surfaces will interfere with the electromagnetic wave produced by radio frequencies or other electrical transmissions. Thin aluminum shields are often used as radio frequency interference (RFI) shields.

Aluminum can be a very ductile metal, allowing shapes of considerable detail to be formed from plates and sheets. It can be rolled into thin ductile foils, cast into highly detailed molds, and economically extruded through intricate cross-sectioned dies. In the hands of a knowledgeable designer, aluminum offers vast possibilities to tailor the base metal to specific criteria.

Corrosion resistance is exceptional, as aluminum develops a tight oxide layer instantaneously when exposed to air. The oxide that develops is very thin and transparent. Aluminum without its oxide is highly reactive due to its desire to shed the three electrons in its outer shell and combine rapidly with other substances. This oxide, Al_2O_3, is a very thin film, 0.25 μm thick. The thickness of the oxide is self-limiting. It grows in thickness quickly, then stops as the thin outer oxide becomes impervious to other elements in the environment. When the oxide is removed by a scratch on the surface, the oxide layer quickly regenerates, making it an excellent metal for coating with paint and lacquer. If a deep scratch occurs through a painted aluminum part, the aluminum oxide will develop and protect the base metal from corrosion (Figure 1.4).

The aluminum oxide is very stable in most exposures, particularly those environments where the pH range is 4–8.5. Seawater is near-neutral pH, making aluminum a good marine environment metal. Seawater does create an excellent electrolyte, so dissimilarity of metals and the associated galvanic corrosion can cause aluminum to be sacrificial to other metals. The oxide is transparent and will not stain other materials; it is also nontoxic to plants and animals (Figure 1.5).

ENVIRONMENTAL AND HYGIENIC

The recycling of aluminum, particularly discarded cans for beverages, has developed into an industry of its own. The collection, shredding, and melting of the metal returns more than 65% of aluminum back into the marketplace each year. Industry recycles all aluminum scrap used in production, keeping it separated by alloy type. In 2013, approximately 50 million tons of aluminum were produced from ore while 20 million tons of aluminum were recycled from scrap material. The recycling of aluminum is an extremely important aspect of the metal production. It effectively keeps costs down and reduces the ravages of mining and production operations on the environment.

Refining aluminum from ore, specifically bauxite, requires massive amounts of energy. The reason large aluminum refinement facilities were built around hydroelectric dams was for the economy and availability of electricity afforded by these generators. Recycling, though, uses only 5% of the energy that is needed to refine the metal from raw ore. Aluminum has a lower melting point than

FIGURE 1.4 The Louvre in Lens, France, designed by SANAA. Unpainted, clear anodized aluminum surface. Metal was bright rolled by AMAG of Austria.

most other metals, making it easy to recycle and to reduce casting costs. Collecting aluminum cans for recycling is a cottage industry for some. Can recycling occurs everywhere in the world, as the high price paid for aluminum scrap provides an incentive to many needing an easy source of revenue.

The introduction of over 100 million tons of carbon a year into the atmosphere from primary production is avoided because of the recycling of aluminum.[2] In the United States, 80% of the

---

[2] Data from the International Aluminum Association.

FIGURE 1.5 Federal Reserve Bank of Boston. Over 40 years of exposure near the sea.
Source: Designed by The Stubbins Associates, Inc.

aluminum produced is from recycled aluminum. The United States has become a leading secondary producer of aluminum due to the emphasis on recycling and the high cost of energy. Only so much recycled aluminum is available, however, and the balance has to be made up from primary production. In each remelting of recycled aluminum, 2% of the metal is lost as some of the aluminum combines with waste products to form dross, which requires significant energy to refine. The lower melting point of aluminum also reduces the energy needed for casting and separation in the process of recycling.

Currently, more than 50% of the world's primary production of aluminum occurs in China, making this country the largest producer of primary aluminum (sometimes referred to as virgin aluminum). Primary aluminum is not from recycled material but from direct smelting from ore.

Aluminum ore is known as bauxite. Bauxite contains significant amounts of alumina, the name given to aluminum oxide. For every 4 k of bauxite mined, 2 k of alumina are produced, and from this 1 k of virgin aluminum is obtained. Bauxite mining is a strip-mining operation, and therefore causes significant environmental damage where it occurs. Strip mining removes the topsoil and the animal habitats that it provides.

Often, the areas where strip mining is undertaken are tropical regions. This is due to the presence of tropical monohydrate bauxite. This grade of bauxite yields high percentages of alumina. The largest bauxite mines are found in Australia, China, Brazil, and Indonesia, while the largest producer of alumina is China.

The refinement process from bauxite can have its environmental consequences. The industry has addressed these and will continue to do so. There are several significant challenges with aluminum production from ore. The primary challenge is the massive amounts of electrical energy needed for the electrolytic process. Large smelting operations consume as much energy in a day as a village would consume in a year. The industry is incorporating renewable sources of energy production, such as solar fields, but there is no good way to get around the massive requirement for electricity to produce the metal. Great strides have been made over the past half-century in reducing the electricity needed to produce aluminum. More needs to be done, but today the processes of aluminum production use nearly 50% less energy than they did decades ago.

The argument can be made that aluminum reduces the use of energy in other areas. The lower density allows for the creation of lightweight transportation vehicles, reducing transportation costs. Aluminum's strength and lightweight nature allow the building of larger airplanes to move more people per unit of energy consumption. The corrosion resistance of the metal benefits the lifetime utility of aluminum. It lasts longer and therefore items made from the metal require less maintenance and fewer replacements, which saves yet more energy. When looking at the larger picture, one could say that a complex balance of energy use and conservation is revealed.

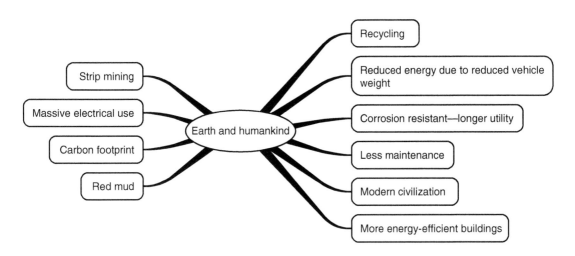

The production of aluminum from the mineral ore source uses cryolite, a fluoride mineral no longer found naturally. It was once mined in Greenland, one of the few places cryolite existed in any sizeable quantity. This source has long since been exhausted and the mineral is extremely rare in its natural form. Cryolite today is made from the more common mineral, fluorite. Cryolite contains sodium and aluminum and has the chemical formula, Na_3AlF_6. During the refining process it is extremely important to capture the fluorine compounds and remove them from the air exhaust from the plant. The release of perfluorocarbon compounds (PFCs) was a major issue in the early years of aluminum production. These compounds are composed of tetrafluoromethane, CF_4, and hexafluoroethane, C_2F_6. Aluminum production is the major source of these compounds and when released into the atmosphere, they remain for an extended period of time. They are found to have a highly detrimental effect on the ozone layer that protects the earth.

The refinement process of aluminum from its ore, bauxite, involves the following stages. Bauxite is a red clay-like sedimentary rock that is mined from the surface and contains large amounts of aluminum in the mineral forms of gibbsite and boehmite, along with other substances. To extract the aluminum the ore is mined, crushed, and mixed with sodium hydroxide in large digesters that put this mixture under heat and pressure.

The sodium hydroxide dissolves the aluminum oxide and forms sodium aluminate. Iron and other heavier impurities sink to the bottom of this mixture and are removed in what is referred to as red mud. Red mud is composed of various mineral byproducts. It is what is left after bauxite digestion, and the red color is from the quantity of iron oxide in the mixture. It has a high water content, which makes it difficult and expensive to transport large distances. It is composed of very fine particles in a liquid slurry. The red mud is produced at a rate of 1–1.5 times that of the aluminum extracted, and seventy to eighty million tons of this waste product are created each year. The alkalinity of the mud is its most serious environmental concern. Although the red mud is not particularly hazardous waste, there is simply a lot of it that is created that needs to be disposed of. The typical way of handling the mud previously was to keep it in large retention ponds for a few years, until the clay could be disposed of. Today there are methods of drying the mud into a less alkaline cake and storing it in stacks. This way it is prepared in a more thoughtful environmental manner and can be more easily transported. It has also found use in road construction, low-cost cement, and in changing the acidity of soils. Recent uses have centered around creating building materials such as tile and brick from the bauxite residue.

Another significant challenge to aluminum smelting has to do with the vast amounts of carbon dioxide released as the carbon anodes are consumed in the electrolytic process. The refinement of aluminum is a continuous process and the carbon electrodes deteriorate over time. As they deteriorate, they release CO_2 into the atmosphere. CO_2 is a greenhouse gas and the aluminum industry recognizes the need to develop technology to control or reverse this development. Today all aluminum production uses rooftop scrubbing equipment to remove the greenhouse gasses from the plant emissions. The processes are monitored to measure the effectiveness of controlling emissions on a plant-by-plant basis around the world. The goal is to remove and eliminate direct greenhouse gas emissions from aluminum production, and much headway has been made.

The oxide of aluminum is not harmful to plant or animal life. Humans come into contact with aluminum every day. Free aluminum rapidly combines with other substances and does not dissolve in water. Aluminum will leach into fluids, however, and affect the taste and flavor of foods and beverages if not coated. Most aluminum cans are coated with a thin film while cooking utensils, pots, and pans are not coated. We are exposed to aluminum mostly through drinking water and cooking as well as from certain cosmetics and some pharmaceuticals. Aluminum has been used for decades in the beverage can industry and for a century in the cooking utensil industry. No detrimental effects from contact with food and oils, even at high temperatures, have been found.

Chemical brightening techniques such as electropolishing and anodizing will remove hydrogen from the surface. These processes bathe the surface in oxygen and thus remove hydrogen. Bacteria cannot multiply on a surface devoid of hydrogen. The brighter surfaces are easier to clean and lack pores that can hold contaminating substances.

Several studies have linked high levels of aluminum exposure with Alzheimer's disease. There is a correlation between elevated levels of aluminum in the brains of people who suffered from Alzheimer's disease, but it is unclear as to whether this is a cause of the disease or the result of the person having the disease. There currently are no definitive links between Alzheimer's and aluminum exposure.

There has been a record of neurotoxicity with people undergoing kidney dialysis. The neurotoxic syndrome is also known as dialysis dementia. The symptoms developed after years of dialysis treatment and elevated levels of aluminum were found in the brain and muscles. The aluminum apparently came from drinking water and could not be filtered out of the body.[3]

There is a lot at stake with the production of aluminum and the industry, as well as the field of environmental research, as scientists are constantly looking at ways to reduce or eliminate the impact of aluminum and aluminum production on the environment. The aluminum industry has published data relating to the environmental impact of all producing areas, which can be found at www.world-aluminium.org. There is a significant driving force behind the development of new production methods in order to alleviate the amount of electrical consumption, the capital cost needed for refinement, and the environmental concerns surrounding the fluoride and greenhouse gas emissions, and the red mud waste products.

HISTORY OF THE METAL

Aluminum was first isolated by the Danish scientist Hans Christian Oersted in 1825. He used a process of isolating aluminum from aluminum chloride that was not simple by any stretch and also not very productive. By using aluminum chloride and amalgamated potassium, a compound of mercury and potassium, he was able to precipitate out small amounts of aluminum. It wasn't until the mid-1800s that a production method for aluminum was developed by the Frenchman Henri Sainte-Claire Deville, using a process where sodium was used instead of the amalgamated potassium. In 1854 Deville devised this chemical process to isolate alumina from bauxite. This chemical refinement process could produce quantities of aluminum but required the use of expensive sodium to reduce chlorine from aluminum chloride. Now with workable quantities of aluminum, this new metal piqued the interest of science and industry due to its amazing qualities, which were never before seen in a metal.

France took the lead in early aluminum production in the nineteenth century. During this time period, just before Deville's aluminum process was devised, aluminum was considered a precious,

---
[3]Gilbert, S.G. (2012). *A Small Dose of Toxicology*, 2e. Healthy World Press.

highly valuable metal, more precious than gold or platinum. It has been reported that Napoleon III would offer aluminum cutlery to his most honored guests, leaving his remaining guests to get by with silver or gold forks and spoons. Napoleon III backed Deville in the manufacture of this bright, strong metal, which he considered of immense value.

Still, even after Deville's manufacturing process proved effective, aluminum was considered novel and precious. The peak of the Washington Monument is adorned with a pyramid of cast aluminum. The largest casting of its time in 1884, this 2.85 kg cast aluminum shape was created by the William Frishmuth foundry in Philadelphia.

During this time, in the United States, aluminum manufacturing was insubstantial: only a total of 28 kg of aluminum were produced in 1885. It was an expensive endeavor and the cost of an ounce of aluminum was equivalent to the cost of silver[4] as the century neared its end (Figure 1.6).

One of the most striking things about the 22.6 cm tall pyramid atop the Washington Monument is the quality of the surface. The removal of trapped gases in castings was not fully understood at the time and the result was often a highly pitted surface. But for this aluminum pyramid, the surface came out of the mold in superior form with minor porosity.

In reality, the use of aluminum as the cap was not as meaningful as it sometimes is made out to be. Certainly, it was a very special material in the late nineteenth century, but it was chosen because

FIGURE 1.6   Aluminum apex of the Washington Monument in Washington, DC.
Source: Theodor Horydczak, photographer.

---

[4]In 1884 aluminum cost $1 an ounce. Demand was very low for such a difficult metal to produce during these early days. Silver cost approximately $1 an ounce. For comparison, the men that worked on the Washington Monument earned $1 a day for a 10-hour shift.

of its electrical conductivity and its nonstaining oxide. The builder of the monument, Colonel Thomas Lincoln Casey, needed a material for a lightning rod. Frishmuth decided on aluminum in lieu of a gold- or platinum-plated copper alloy. They wanted a material that would not stain the stone masonry below. They feared the plating would eventually wear away and expose the copper to weathering, and the resulting green patina would leach onto the monument's stone. This new metal, aluminum, met all the criteria as a capping point to this most important monument. Plus, Frishmuth owned the foundry that made the small pyramid. Most likely, it would still today be the metal of choice for the tip of the monument.

THE ORIGINS OF MODERN PRODUCTION

Not long after this early casting, in the year 1886, two people in separate parts of the globe discovered a process of electrolytic reduction to wrest aluminum from its ore. Paul Héroult in France and Charles Hall in the United States nearly simultaneously discovered the electrolysis process that is still in use. What they discovered ushered in a process that opened the metal for commercial use in industry and architecture.

Both Paul Héroult and Charles Hall were only 23 years old when they isolated aluminum by running an electrical current through a solution containing dissolved aluminum compounds. Silvery nodules of pure aluminum precipitated out at the bottom of the beaker. The solution they used was molten cryolite, which dissolves aluminum oxide. Running a strong electrical current caused the aluminum to come out of solution and deposit on the electrode. Hall used a series of batteries while Héroult used an electric dynamo that his mother financed with her life savings. The need for electrical energy is still a demanding requirement for the production of aluminum. Hall worked closely with his older sister, Julia Brainerd Hall. She took impeccable notes of the work that later became part of the proof that Hall's discovery was a few weeks ahead of Paul Héroult's. Subsequent patent disputes between the two ensued but eventually they came to mutual terms and both successfully profited from the Hall-Héroult Process, as it came to be known. Only once in their lifetimes did the two men meet in person. Born in the same year, Hall and Héroult also died in the same year, 1914.

In the United States, Hall joined up with the Pittsburg Reduction Company, later renamed Aluminum Company of America, or Alcoa, to mass produce aluminum. Meanwhile, Héroult sold his patent to Aluminium Industrie Aktien Gesellschaft in Switzerland, which soon because Alusuisse-Lonza. These two companies controlled the production of aluminum and hired designers to come up with new, unique uses of the metal.

Within just two years, Hall's company was able to produce 50 pounds of aluminum a day. By the early 1900s Alcoa was producing more than 80,000 pounds a day. As the price of the metal came down, it lost its novelty as a metal for the wealthy and became one of the primary industrial and architectural metals of the modern age.

Since the beginning of the twentieth century, innovations in alloying and production processes significantly enhanced the marketplace for this special metal. Casting processes improved and allowed for larger production parts. In the 1930s, direct chill casting was one of the most notable early innovations in the process of making the raw material for industrial use. Before this, tilt molds made out of steel were used. These were in use in the production of other metals, but with aluminum the steel molds had a detrimental effect on the quality of the metal and created a lot of porosity and dross. William Ennor of Alcoa devised a way of applying water directly to the aluminum ingot as it was cast. The molten aluminum would be poured or dropped by gravity and water would be applied to the aluminum as it descended. The outside would rapidly cool while the interior of the billet cooled more slowly. This allowed for larger ingots to be cast and it eliminated the porosity created by pouring into a tilt mold made of steel. Larger castings were now possible and larger ingots could be created for industrial use.

The Empire State Building, designed by Shreve, Lamb, and Harmon and completed in 1931, is adorned with numerous large cast aluminum exterior panels. These art deco castings, some of the largest in their day, were cast from aluminum alloy 195 (Unified Numbering System [UNS] alloy A12950), a copper and silicon alloy.

In the United States, early uses of aluminum centered around the decorative art castings for buildings like the Empire State Building and the Chrysler Building. This new fascinating metal was lightweight and would not corrode like steel castings or stain surfaces like bronze castings.

Three major buildings in Chicago—the Venetian, the Isabella, and the Monadnock—used aluminum as artistic surfaces and embellishments in the late 1800s. In 1926 the artist Henry Hornbostel designed and cast an elaborate aluminum spire for the Smithfield Church in Pittsburgh (Figure 1.7).

The drawback to these early castings was the lack of detail that could be achieved in the large cast panels—steel castings could achieve far more detail. Casting of steel and iron was in full form in the late nineteenth century and early twentieth century. Casting aluminum required less energy but achieving a good surface was difficult and getting the metal to flow into areas of detail without solidifying was not easy. Therefore early in its history, aluminum was delegated to smaller decorative castings or more utilitarian items such as cooking utensils or tableware.

Aluminum production coincided with one of the major advancements in technology: air flight. It was the advent of aeronautics that pushed aluminum and aluminum development beyond the simple and novel. Aeronautics required sound castings for lightweight engines. The Wright Brothers, seeking a lightweight but strong material, created their engine from this new metal. As the fledgling aerospace industry grew, so did aluminum. Aluminum provided the answer to the search for a lightweight yet durable material that could be fashioned into a small yet powerful engine. The aerospace industry would not be in existence if it was not for aluminum. The world took note.

Airships needed the fiber strength achieved with rolled plates and sheets of aluminum. These became the strong, lightweight skins for aircraft.

Unfortunately, throughout mankind's history, it has often been wars that provided funding and sped up material science development and production capacity to new limits.

18 Chapter 1 Introduction to Aluminum (Aluminium)

FIGURE 1.7 Cast aluminum spire on the Smithfield Church.
Source: Close-up photograph courtesy of Richard Pfiefer, taken when he inspected the steeple's condition after decades of exposure.

Alloying of metals to achieve certain, specific properties was understood by metallurgists, so early development of aluminum centered around alloying (Figure 1.8).

Much of the early development of the alloying of aluminum occurred in Germany. In 1906, Dr. Alfred Wilm of Germany experimented with alloying aluminum and different treatment processes from those that were used on other metals. During one of his experiments, he created an alloy of aluminum with measures of copper and magnesium. He heat-treated the alloy by heating to a specific temperature and then cooling the metal in ways similar to heat treatment processes used on steels at the time. He then measured the strength and hardness. The results were not very good. The aluminum was weak and soft, so Dr. Wilm set it off to the side and carried on with his other experiments. After a few days, for some reason he went back to recheck his findings and measured the sample a second time. What he found was that the aluminum alloy had gained in strength and hardness. At the time, he had no idea what was occurring, but his repeated tests arrived at the same results: an increase of strength and hardness with time. This was the discovery of a phenomena known as

FIGURE 1.8 Image of the vast aluminum skeleton framing of the Hindenburg zeppelin and the moment of disaster when the airship's hydrogen gas ignited.

"age hardening," which is a property of the more malleable metals and alloys that contain titanium, magnesium, and nickel. The alloy became known as "Duralumin" and is similar to the alloy A92017 that is in use today. The process became what is called "solution heat treatment and naturally aged." Duralumin was developed in Germany just prior to World War I as a lightweight metal to be used for the early "airships," more commonly referred to as blimps. Small amounts of copper were added to aluminum as the alloying constituent and it was found that strength could be enhanced significantly. Duralumin is a strong aluminum alloy, and was stronger than the aluminum used up to that time. The process enabled aluminum to enter the Industrial Age and led to the discovery of a number of strengthening processes for the metal.

One significant problem with Duralumin was corrosion. This alloy had poor corrosion resistance in salt environments. It was found that by cladding the alloy with a purer alloy of aluminum, it would be significantly more corrosion resistant without losing any of its strength characteristics. This spawned the process known as alclad. Alclad became a common procedure for corrosion protection of the high-strength alloys.

As the world wars took place, all aluminum production was centered around military use of the metal. Various alloys and processes of strengthening them would create an entirely new vocabulary for aluminum.

It was in the first few decades of the twentieth century that research into the possibilities of aluminum and other nonferrous metals began to take shape. In England the Vickers Corporation, a British armament firm, was tasked with studying aluminum for use in ship construction and later in "air ships," to compete with the dirigibles in use by the Germans. The British, French, and Germans had more advanced metallurgical research into the fundamentals of aluminum in these early years. When World War I broke out, much of this research centered around armaments and ships. The aluminum alloys known at the time were not sufficiently strong and were prone to failure from corrosion. Airplanes were still made mostly of wood and canvas, materials that were better understood than this new lightweight metal.

In the United States, serious research did not take off until around 1914. Casting was the main fabrication process and was performed in steel tilt molds, but the quality was poor and unpredictable. Porosity from entrapped hydrogen and shrinkage cracking limited the use of the metal. In 1919 Alcoa, combined applied research and development into one technical department and this enabled the metals capabilities to expand. After this restructuring, developments with aluminum as a serious industrial material unfolded. In 1923, the horizontal press using preheated billets of aluminum advanced the available forms of extrusion. Hydraulic extruding of metals such as lead and copper was underway long before the arrival of aluminum, but advances in techniques and strengthening processes allowed the extruding of aluminum shapes into a vast array of cross sections. The extrusion press was an innovation utilized in aluminum alloy fabrication and advanced design. The hydraulic extrusion could produce long lengths of designed forms by pushing the heated metal through a steel die. Aluminum extrusion came into use soon after the refinement process reduced its cost and increased its utility. The G.A. Dick Corporation extruded aluminum in 1884 using a process developed for copper and zinc extrusion. Aluminum was ideal for extruding. It did not require high pressures to push it through the extrusion die. The trouble, at first, was that heating the aluminum to push it through the press also weakened it. A process known as "press quenching" was developed to rapidly cool the metal. This eliminated a secondary heat treatment process. The early press quenching processes started by using fire hoses to cool the metal down.

New casting methods, such as the Direct Chill Process developed by William Ennor in 1930, greatly enhanced the efficiency of casting ingots, which further enhanced the mechanical properties available with aluminum. New, stronger alloys were developed from these high-quality ingots, which could now be produced in quantity (Figure 1.9).

Aluminum was an ideal material for another new invention, the electric light. As a reflector, aluminum could reflect light and enhance a glow without altering the color. Oxide and corrosion reduced some of the benefit that these early reflectors provided, but compared with other materials at the time, aluminum's reflective properties were superior.

Anodizing was introduced in the United States in the 1920s. The sulfuric acid electrolyte, still used today, was introduced in the 1930s. Anodizing helped with the corrosion issues but dulled the reflection. This led to the development of brightening practices added to the anodizing process. Now aluminum would have improved corrosion resistance *and* a highly reflective surface.

During World War II, aluminum production expanded dramatically. As high strength alloys were developed, and marine grade alloys were refined, aluminum became a critical military material. In the United States, Alcoa could not supply all the aluminum the war effort required, so the government set up production in several facilities. Up until the 1940s Alcoa had a monopoly on the production of aluminum in the United States. Reynolds Aluminum and Kaiser Aluminum

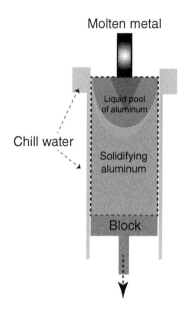

FIGURE 1.9 Direct Chill Process of aluminum production.

joined in the production of the metal, greatly increasing the number of alloys in use and the methods of production. These plants were run by Alcoa but financed by the US government. During World War II, with the military advancement of airplanes playing a significant part in warfare, massive amounts of aluminum were needed. Steel was in short supply and aluminum production was ramped up. Extrusion processes, rolling mills, and bauxite refineries were set up. Special alloys, such as A96063, were introduced in 1944. This alloy, like its cousin, A96061, had good strength, could be heat treated, and retained excellent corrosion resistance. The A96063 alloy could be produced quickly for war production.

High-strength alloys such as A92024 were created for aircraft. These were significantly stronger than earlier alloys. Alloy A92024 possessed a design strength of 57,000 psi. This allowed the first commercial airliner in 1935, the DC-3, to be successfully built and flown. By 1943, the United States was producing 835,000 tons of aluminum. By the end of World War II, the United States had produced more than 300,000 airplanes from alloy A92024. Near the end of the war, another high-strength alloy A97075 was produced and used to create the Boeing B29 Bomber. This alloy, containing 5.5% zinc, had a design strength of 73,000 psi. Aluminum was no longer just a novel, lightweight, decorative metal. It was now challenging steel as a metal chosen for strength.

Shipbuilding also saw the benefits of aluminum. Aluminum alloys containing magnesium were found to be more resistant to marine environments than other aluminum alloys and significantly more resistant than steels. The alloys A95083 and A95086 were developed for the maritime industry.

Aluminum's industrial use continued to accelerate after the World War II, mainly due to the capacity generated from the war effort and the subsequent drop in production cost.

Alcoa's monopoly ended with the introduction of two new companies: Reynolds and Kaiser Aluminum. After the war, production capacity was turned over to the private sector and new uses of the metal in industry expanded rapidly. Today, flat-rolling, extrusion, tubing, and casting of aluminum are commonplace in the building and construction industry. Near the end of the war, alloy A96063 was developed for extrusions. This alloy has excellent strength, good corrosion resistance, and could be anodized both clear and with color. Alloy A96063 is now the most common extruded alloy used in art and architecture today.

ALUMINUM IN ART

Aluminum has a rich history in art, but its progress happened over many years. Aluminum could be more readily hammered and machined than steels and some copper alloys. Casting temperatures are lower, and its softness made it ideal for shaping and finishing. Aluminum's ability to maintain its color and tone and resist tarnish made it ideal for many decorative small pieces and intricate shapes, such as gears for timepieces.

In the mid-1800s aluminum was an expensive metal. Early uses included helmets worn during parades by aristocracy, such as Ferdinand, Hereditary Prince of Denmark. This silver-white helmet decorated with gold was lightweight compared to other metals and resisted tarnish, unlike a silver-plated steel helmet.

As the metal became more affordable, firefighters, construction workers, and others who needed a high level of protection adopted the use of aluminum for their helmets. Its lightweight nature and the way it kept the head cool were the attributes in demand (Figure 1.10).

When the production process changed after the development of the Hall-Héroult process, the price dropped significantly, and the metal was more available to industry and design. Its use as a design material expanded mainly in Europe, with the French leading early development. Artists and designers used the metal as cast and hammered forms desirable for both its color and ability to resist tarnish and corrosion.

The Art Deco style, born out of the *Arts Decoratifs et Industriels Modernes* exhibition in France in the early 1900s, embraced this new metal with its silver color and ability to be shaped and formed into sleek curved surfaces. Furniture makers and art deco architects soon found ways of using this new metal, which would not tarnish like silver and could be easily formed into decorative objects. In France during the early part of the twentieth century there was a strong art deco movement that incorporated aluminum in artistic forms depicting a new modern age. Aluminum and its ductility along with the ease of casting and polishing allowed the designer to create curving, flowing forms. The surface of the aluminum could be anodized, giving it an easy-to-clean and maintain surface.

Soon this art style was imported to the United States and aluminum was woven into many Art Deco designs from household wares to storefronts.

Early French designs involved aluminum with contrasting brass overlays. Hammered and etched textures could be imparted to the aluminum to generate the designs we associate with the Art Deco style (Figure 1.11).

Aluminum in Art   23

FIGURE 1.10 Setting the lightweight aluminum sculpture by helicopter.
Source: Ron Fischer, artist.

FIGURE 1.11 Examples of art deco artwork.

This style was imported to the United States and embraced by both architects and designers. The Kensington Company was one of the most prolific companies to produce art deco artwork with aluminum. Kensington was a brand developed by Alcoa to promote the use of aluminum in households across America. As Alcoa sought new avenues for this new metal, they opened subsidiaries to produce and promote the metal. In 1901, some of the first products were cooking utensils for the home. A subsidiary known as The Aluminum Cooking Utensil Co., Inc., was formed. In 1903 it changed its name to the Wear-Ever Company.

When working with aluminum, artists should recognize certain properties that are inherent to the metal. The metal can be cast, welded, cut, and formed in ways that are similar to other metals. Its corrosion resistance, attractive color, and ability to reflect over 90% of the visible light spectrum bestows aluminum with a special metallic feel that often is mistaken for chrome. The luster of aluminum is deeper than chrome and comparable with silver. When aluminum is polished to a high luster by hand, generating the subtle highs and lows of human inaccuracies in pressure, then chemically oxidized with a darkened tone, it takes on the appearance of wrought silver (Figure 1.12).

The ease of shaping and polishing aluminum and the ability of the metal to resist tarnish enabled the artist to capitalize on this new metal. It is essential to understand the properties of the metal and the limitations of the material. Aluminum is different in many ways from other metals used in art and architecture. These differences can offer profound attributes to the designer and

FIGURE 1.12 Polished and smoothed mirror aluminum sculpture.
Source: Jan Hendrix, artist.

enable the sculpting of remarkably lightweight shapes that will withstand environmental exposure and the test of time.

PRODUCTION PROCESS TODAY

The production of aluminum has evolved into two major sources of base material: the primary source of newly refined aluminum or aluminum produced from the ore, and the secondary source, aluminum produced from recycling. North America and Europe produce much of the aluminum from recycled stock and purchased ingot from other countries. China produces the greatest amount of primary aluminum today—more than 50% of the world's supply comes from this region. Other areas of the world, especially places where energy is more economical, are also developing primary aluminum production. Hydroelectric sources or natural gas sources, for example, have become the chosen centers for primary production. Qatar, for example, has an abundance of natural gas, which supports the operation of a large modern smelting operation producing primary aluminum from ore.

Primary aluminum production centers around the smelting of alumina into pure and highly pure forms of aluminum. Pure aluminum is considered 99.0–99.9% purity. Traces of other elements, mostly iron and silicon from the bauxite, remain. Highly pure forms of aluminum are 99.97% pure. These are needed for the electrical industry. These forms are produced by subjecting the aluminum to further refinement to remove traces of elements such as titanium, vanadium, and chromium, which have a detrimental effect on electrical conductivity.

These pure forms are the basis for further alloying processes performed at the Mill. The higher purity forms are combined with various prescribed amounts of alloying elements and secondary aluminum from scrap to arrive at the specific aluminum alloy needed to fulfill a specified order.

Secondary aluminum production is the refinement of scrap. Scrap is a general term that includes:

- Processing scrap left over from specific fabrication operations
- General scrap from old recycled end-of-life products
- UBC—Used beverage cans
- Dross—Residual from cast process

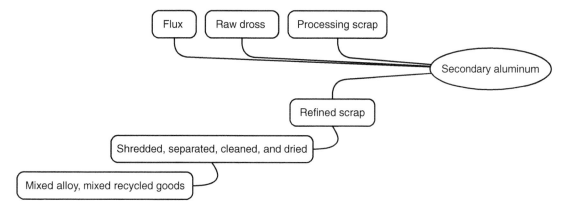

Secondary aluminum production has become more specialized in the identification and separation into the specific alloy type. There are specific plants today that specialize in the recycling of the UBC. Here it is critical to remove and separate the lacquers and organic matter and concentrate the recovery of the greatest amount of aluminum. Additionally, end-of-life scrap from painted building products and household goods require separation and differentiation from the coatings and organic material that may accompany them into the recycled bin.

The most valuable scrap is the processing leftover and collected at the facility making the end products. Here the alloy is known. Collected at the factory from milling, cutting, and stamping processes along with extrusion ends and billet ends, these scrap items, defined by their original order and verified by XRF[5] or other devices, obtain premium prices and are in more demand than mixed aluminum scrap (Figure 1.13).

The aluminum mill takes the pure forms and secondary forms of the metal and produces the raw forms of the material used to make architectural wrought finished and semi-finished products, as well as the alloy cast stock for the aluminum foundry. Initially, the Mill creates what is called a heat. A heat involves melting various sources of aluminum, such as scrap and high purity blocks called pigs, adds in specific alloying elements, and generates a casting of a specific alloy to recognized

FIGURE 1.13 Image of XRF reading of aluminum alloys.

---

[5]XRF: X-ray fluorescence. A handheld device used to bombard the surface of a material with X-rays and read the trace elements that make up the material surface.

tolerances of impurities. This heat casting is large, several thousand tons of aluminum. Depending on the order, the heat is cast into one of three forms: large rectangular slabs, large rods, or smaller rectilinear shaped blocks. These forms are precursors to the next refinement of the metal.

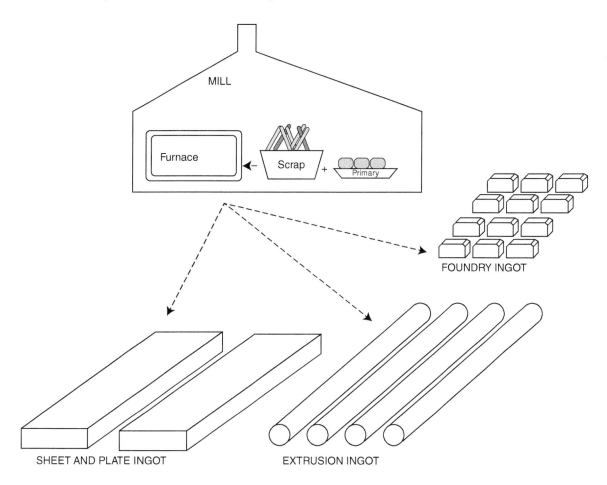

The forms destined to be turned into wrought aluminum (plate, sheet, wire, tube rod, or extrusion) are referred to as ingots. For plate and sheet wrought forms, these ingots are large rectangular slabs weighing several tons.

The cylindrical rod forms are ingots to be used in extrusion presses, pipe and tube manufacturing, and wire manufacturing. The smaller blocks are set for specific casting operations and are referred to as mold casting ingots (Figure 1.14).

When the aluminum is cast into the various forms, the impurities and oxides appear on the top portion as they float on the molten metal. These are removed in a process called scalping.

FIGURE 1.14 Ingots for extrusion and foundry.

This trims the various blocks to a set size and the scalped metal goes back into recycling to recover as much of the aluminum as possible. This limits the impurities from becoming part of the more refined form.

The Mill takes the large, scalped slabs and puts them through a hot reducing operation similar to what occurs at a steel mill or copper mill. The metal is heated to between 350 and 550°C and passed through a series of rolls. The rolls put pressure on the hot slab of metal and squeeze it to reduce the thickness and widen and lengthen the slab until it becomes a long ribbon of thick plate. Further rolling is performed on cold rolls that squeeze the metal and lengthen the ribbon. A temper is imparted into the sheet or plate material and develops the refined finish on the surface of these forms. Annealing and stretcher leveling of the metal occurs at this time. The semi-finished material is coiled for further refining or finishing.

The cylindrical ingot also is scalped where the outer surface is trimmed and the ends are cut into a billet size to be used at an extrusion plant where it will be drawn, stretched, and straightened into a predesigned custom extrusion shape, rod, tube, pipe, or wire.

Other ingots are designed and set out for a forging facility where the metal is hot-forged into a particular shape or form. These ingots are sized for the forge plant. The ingots set for casting are sent to sand, die, and permanent mold-casting operations, and other specialized casting facilities.

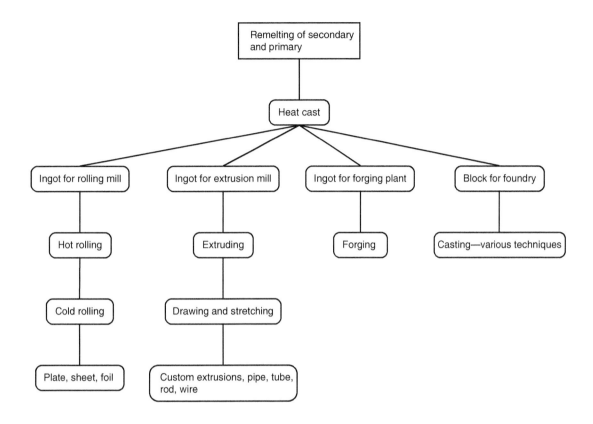

ALUMINUM VERSUS ALUMINIUM

Sometime around 1807, Sir Humphrey Davy was busy discovering elements like potassium, sodium, and magnesium. He used electrolysis to isolate metals from molten compounds. He did not isolate element 13, but he knew it must exist. He first spelled the name as alumium. A few years later, he changed the name to aluminum. Then again, in 1812, he called it aluminium and this name stuck, at least in Europe.

After Charles Martin Hall discovered the process of producing and manufacturing aluminium and started his company Alcoa in America, he needed to advertise and market the metal. In his early publications he dropped the "i" and called his metal aluminum, and the name stuck in American jargon. Everywhere else in the world the metal goes by the name "aluminium."[6]

---

[6] Author's note: By using the American spelling of the word aluminum instead of aluminium, I have kept from pressing the "I" key many times. Think of the wear on the "I" keys and the ink Mr. Hall has saved with his efficient removal of the self-important "I."

COMPARISONS BETWEEN ARCHITECTURAL METALS

Comparison of various metal attributes in a general sense is not a simple endeavor because all metals are alloys and the alloying aspects can markedly alter the performance characteristics. For example, aluminum can be milled to create significant detail with ease. Copper alloys can hold detail as well, if there is a little lead present in the mix. Steels are notoriously poor when it comes to corrosion resistance until you add a little copper and chromium to the alloy to make weathering steel, or coat them in zinc via the process of galvanizing. The treatments for each metal, and the way their respective industries have worked to overcome challenges, have greatly increased the utility of the world of metals. Table 1.2 provides a general comparison and selectively identifies a few of the alloys to indicate in particular the metals used in art and architecture.

Generally, in art and architecture, the way that a metal will appear is a high priority. Some considerations are the color and reflectivity and how the metal's surface will change with weathering.

TABLE 1.2 Color and appearance.

Metal	Natural color	Color potential	Expected change in color	Reflectivity
Aluminum alloys	Silver white	Vast with dyes	Some dulling over time from UV and oxide development	Specular and nonspecular
Stainless steel alloys	Silver	Vast with physical vapor deposition (PVD) and interference colors	Long life in color and tone. Some colors decay in certain exposures.	Specular and nonspecular
Copper alloys	Red brown	Vast with alloys and patinas	Long life in color and tone with certain patinas.	Nonspecular unless protected with coating
Zinc	Gray	Moderate with patina	Long life in color and tone with certain patinas and exposures.	Nonspecular
Steel—cold rolled alloys	Dark gray	Limited to dark tones	Moderate lifespan when coated. Best for interior uses.	Nonspecular
Steel—weathering	Red to deep brown	Limited	Long life span in most environments	Nonspecular
Steel—hot dipped galvanized	Gray—silver	Limited. Can darken.	Moderate lifespan.	Nonspecular
Titanium	Golden silver	Vast with interference coloring	Long life span in most exposures.	Nonspecular

The forms and fabrication ability along with strength and cost will play a role in a particular metal selection. Certain metals are not readily available in various forms and limitations on the skill of the factories will influence cost and end quality of the work (see Table 1.3).

TABLE 1.3 Forms and ease of fabrication.

Metal	Variety of available forms	Mechanical strength	Ease of working with forms	Relative cost
Aluminum alloys	Vast. All forms. Extrusions are one of the most significant design form.	Depending on the alloy and temper, good strength characteristics.	Malleable and ductile. Welding requires knowledge and skill. Limited solder and brazing. Casting requires skill.	Low to moderate
Stainless steel alloys	Coil, sheet and plate, rod, tube, and pipe, as well as wire. Limited casting.	Very good strength characteristics.	Malleable but work hardens quickly. Need skill and knowledge to work metal. Excellent material to weld. Solder and brazing require skill. Casting limited.	Moderate
Copper alloys	Foil, sheet, rod, tube, and pipe. Limited extrusions. Rod and wires. Castings.	Relatively low to medium strength characteristics. Improve with alloys.	Malleable and ductile. Weldable, solders and brazes with ease. Alloy casting is common.	Moderate to high
Zinc	Sheet, plate, rod, limited extrusions. Casting limited.	Relatively low strength.	Malleable and ductile at room temperatures. Weldable, but limited. Castable.	Moderate
Steel—cold rolled alloys	Sheet, plate, rod, tube, pipe, and wire.	Good to excellent strength.	Malleable and ductile. Formable, weldable, and castable.	Low
Steel—weathering	Thick sheet and plate. Bar, rod, pipe, and tube.	Excellent strength.	Formable but requires more power. Weldable and castable.	Low
Steel—hot dipped galvanized	Sheet, plate, rod, tube, pipe, and wire.	Excellent strength.	Malleable and ductile. Formable. Welding will damage coating.	Low
Titanium	Thin sheet	Excellent strength.	Formable in certain tempers. Can be welded with special skill.	Very high

CHAPTER 2

Aluminum Alloys

"In that direction," the Cat said, waving its right paw round, "lives a Hatter: and in that direction," waving the other paw, "lives a March Hare. Visit either you like: they're both mad."

—Lewis Carroll, *Alice's Adventures in Wonderland*

In a way, aluminum is not just one metal, but several. This chapter discusses various alloys that make up the metal we refer to as aluminum. The characteristics of the different alloys of aluminum are defined by small amounts of additive elements. Slight variations in these additive elements create significant alterations in the mechanical properties and corrosion behavior of the aluminum alloy. There is no other metal for which just minor changes in the alloying materials will have such significant ramifications to the performance and strength properties. Because of this, it is very important to the designer to understand the differences and work with the manufacturer and producer of the metal to arrive at the optimum available alloy for their structure's particular environment and expected service.

The aluminum industry has defined eight distinctive classifications of the metal aluminum (Table 2.1). With stainless steels, there are four major classifications, and to alter the properties of stainless steels requires significant differences in alloying elements. There are numerous copper alloys, but any significant modifications involve major differences in the amount of copper. Aluminum, however, responds to minute differences that effectively alter properties of formability, strength, and corrosion resistance. As little as 1% of other elements is all that is needed to effectively alter the metal from one aluminum alloy to the next.

The importance of aligning the distinctive aluminum alloy with its end use can have profound consequences on the long-term success of the application. Some alloys perform better in marine

TABLE 2.1 Classifications of aluminum wrought alloys.

A91xxx series	99% aluminum	Trace amounts of silicon, iron and other elements
A92xxx series	Copper	Copper improves strength; traces of other elements
A93xxx series	Manganese	Trace amounts of other elements as well
A94xxx series	Silicon	4.5–6% silicon along with trace amounts of other elements
A95xxx series	Magnesium	Small amounts of magnesium along with a few trace elements
A96xxx series	Magnesium and silicon	Both elements added in controlled amounts, maximum 2%
A97xxx series	Zinc	Added for strength; magnesium and copper as well
A98xxx series	Other elements	

environments and others will produce better surface clarity and appearance when anodized. Certain alloys can be strengthened by heat treatments to improve the mechanical strength. It is these alloying constituents mixed with the aluminum that determine aesthetic, corrosion resistance, and mechanical behavior.

There are currently several hundred alloys of aluminum. Most were created, refined, and perfected for a particular industry. The development of special alloys of aluminum is an ongoing process.

Lithium, the lightest metal, is showing promise in the aerospace industry because when alloyed with aluminum it creates a lightweight but very strong aluminum alloy. Currently it is not used in art and architecture due to its higher cost and lower corrosion resistance. Additions of small quantities of lithium, up to 2.7%, can improve the stiffness of aluminum and at the same time reduce the density. One alloy, A98090, contains from 2.2% to 2.7% lithium along with small amounts of copper and magnesium. It has a yield strength of 370 Mpa (53,700 ksi). Used for defense weaponry and advanced aeronautics, this highly specialized alloy has a density that is nearly 6% less that of other aluminum alloys.

In recent years scandium, a rare earth metal, has been incorporated into select alloys of aluminum. Scandium additions in very small amounts increase tensile and yield strengths. This very expensive element produces a very refined grain when alloyed in aluminum. Aluminum-scandium alloys are used in special high-end sporting goods as well as the Russian MiG fighter jet.

Improving the properties of aluminum by alloying element additions and special thermal processes is ongoing. For architecture and the art market, the surface characteristics and quality are paramount for both visual clarity and predictability. An understanding of the behaviors created by the alloying elements is crucial.

Aluminum is not just one metal, but several. The characteristics of the different alloys of aluminum are defined by small amounts of additive elements. Slight variations in these additive elements create significant alterations in the mechanical properties and corrosion behavior of the aluminum alloy.

CHOOSING THE CORRECT ALLOY

Depending on the quantity, the choice of alloy may be limited by the producer of the form and finish. Many producers of prefinished goods and stocking warehouses carry only those alloys and the tempers in demand. If the quantity is significant enough, you can usually work with a manufacturer and order the tailored alloy and temper for your precise requirements.

For art and architecture there are several common alloys that are often available in smaller quantities. In the wrought forms of sheet, plate, or extrusion these would be the A95xxxx and the A96xxxx. If the aluminum is to be anodized, the anodize quality (AQ) designation is important. AQ differs from the designation of "commercial quality" by having a superior, more refined surface. The A95xxx alloys come in AQ.

There are other alloys that are best suited for extruding. The A96061 and A96063 alloys are common alloys stocked at many extrusion press operations. Many tube forms are available in one of these alloys. The temper is usually a T6 temper with slight modifications.

If the project is near the seacoast, there are alloys designed to perform well when exposed to the chlorides that are prevalent in coastal regions. These are the A95083, A95086, and A95754 alloys. These alloys can be finished by anodizing to give further protection.

Forming operations—in particular the specialized forming operations of superplastic, deep drawing, and chemical milling—have more refined alloying requirements. Welding operations where structural integrity must be considered will have more specific requirements on the alloy of aluminum.

The following section is a guide to the alloys considered for architect and art. It is advised to work closely with the supplier of the metal to arrive at the best choice for the particular purpose.

THE INITIAL MILL CASTING

Understanding the limitations and constraints That the aluminum mill must overcome during the initial mill casting is important. It is not as simple as making a cake, unless the cake is a 20 000 k cake. We often hear the metal must come from the same "heat" in order to arrive at a uniform surface appearance with consistent mechanical properties. The initial heat casting is where this all begins.

At the Mill, the initial ingot of metal is created from high purity aluminum blocks called sows and pigs. To these are added additions of scrap aluminum that has been cleaned and heated and a few additional elements included by metallurgic design to enable certain mechanical attributes and improve corrosion resistance.

The first step in the process of creating the heat is charging the furnace. Here the prime metal is added in one of three forms: t-bars; tub sows, if greater than 25 kg; and pigs, if less than 25 kg. These blocks of aluminum are 99% pure and are referred to as "primes." There are also 10/20 primes, which contain 0.10% silicon and 0.20% iron, and are still considered very pure. After this charge, scrap is added. The scrap is not arbitrary but is sorted and qualified by alloying makeup. Other elements are added to the furnace to produce desired properties in the heat and the mechanical properties needed for the end product.

The furnace initially is brought up to a high temperature to remove moisture. Additionally, there is always what is referred to as a "heal" remaining in the furnace from the previous heat. A heal is what coats the sides of the furnace. The weight of the heal determines how much new prime will be added.

This collection of aluminum is what makes up a heat. The total weight is anywhere from 9,000 to approximately 32,000 kg, depending on the furnace.

As this collection of aluminum is melted, a step called "degassing" is performed in which the hydrogen is removed. Molten aluminum absorbs hydrogen as it combines with the oxygen in H_2O to readily form its oxide Al_2O_3, leaving the hydrogen dissolved as a gas. As the aluminum solidifies, the hydrogen is released, but it is desirable to pull the hydrogen out to reduce porosity in the cast. Gases are pumped into the molten metal along with other fluxes to remove these unwanted impurities. The impurities rise to the surface and form what is called "dross." At this stage a sample of the molten metal is taken, and the chemistry is verified to ensure that the proper alloying constituents fall within the specified ranges.

The next step is to remove the dross, which is a mixture of aluminum oxides and other impurities. These are not discarded but are recycled to remove as much of the aluminum as possible. The longer the aluminum remains in the molten state, the more dross develops as moisture is absorbed.

Once solidification occurs, another analysis is performed with a spectrometer to arrive at the precise alloying makeup of the cast block of aluminum. This is part of the mill certification of this particular heat of metal (Figure 2.1).

Following this is the final step of the heat production. This step is called the "drop." The metal is poured into a trough and chilled using the direct chill method. This creates a series of large blocks of alloyed metal in forms and shapes to be further refined into the mill product. This could be rectangular blocks for creating plate or coils for sheets. This could be round billets for extruding processes or blocks for creating wire. All these initial blocks are from the single heat and near identical alloying makeup.

The next heat, even if the alloy is specified to be the same, will have slightly differing makeup. This is the first constraint to the process. As with stone cut from a mountain side, or wood from a series of similar tree species, the variations are natural and cannot be overcome. The alloying

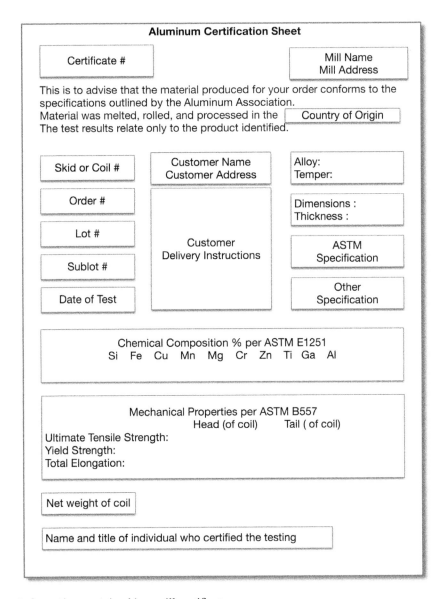

FIGURE 2.1 Information contained in a mill certificate.

makeup is within the standards of control established by decades of industrial research and production. There always are ranges within a given alloy that establish what a mill can meet. These have been improved by industry over the years but trying to achieve tighter requirements has diminishing returns, in both cost and practicality. The mechanical properties will be very close, but the appearance may show differences. For art and architectural use, this comes into play on certain anodized finishes and certain mill finishes. These finishes can display slight variations due to the minor metallurgic variations that different heats possess. Certain designs may exceed a given heat size limitation. Supply of design elements may come at different times and thus arrive from different heats due to timing, supply parameters, or because of damages to installed work.

There are hundreds of alloys of aluminum developed for general and specific purposes. Still more are in development. Of the eight classifications of aluminum alloys, for the world of architecture, the majority of applications use an alloy of one of three categories. Alloys in the categories A93xxx, A95xxx, and A96xxx make up the greatest share of this segment of the architecture and art marketplace. An occasional clad (Alclad) alloy in the A91xxx classification is used to improve corrosion resistance by coating the surface of one of the less corrosion-resistant alloys, but for the most part, A93xxx, A95xxx, and A96xxx make up the majority of aluminum alloys used in architecture and art.

In this section, the various alloys that find their way into architecture and art are described in more detail along with several that reside on the edge of convention. Common tempers are included for many of these alloys and the approximate mechanical properties that these tempers furnish are listed. Not all the tempers available for a particular alloy are listed. Several commonly used tempers are described along with the associated mechanical properties. Specific mechanical properties should be obtained from the producers of the aluminum along with the test verifications from the mill once the alloys are produced.

COMMERCIAL PURE ALUMINUM ALLOY A91XXX

Pure forms of aluminum are very corrosion resistant. They are also soft and weak, regardless of how much cold working is performed. In commercial use, pure aluminum is not used. The alloys of the A91xxx series are the purest aluminum used. They are considered "commercially pure" but they still have trace elements added to improve strength and because producing 100% pure aluminum is difficult and expensive.

Commercial pure alloys are used as a coating on other metals or other aluminum alloys, when reflectivity and corrosion resistance are paramount. Another commercial use for pure alloys is when ease of forming and shaping is desired, and strength is not critical. In the case of a coating they can be applied by dipping or spraying over other metals including other aluminum alloys. Sometimes referred to as "alclad," this is a process of coating one alloy of aluminum with another purer alloy to improve corrosion resistance. This thin layer protects the core metal and gives it a higher reflective surface. However, the soft outer layer is more susceptible to scratching and wear.

Commercial pure aluminum is weak in a mechanical sense. It has a low tensile and yield strength. It is ductile and malleable and only marginally work-hardens. Wire used for electrical conductivity is often in the higher commercial purity levels due to the improved electrical conductivity. When woven into a braided cable, the strength is improved by bundling the strands.

All aluminum alloys in commercial use contain small amounts of copper, silicon, zinc, and magnesium. There are other elements that are added to impart desirable characteristics into the aluminum, usually with tradeoffs in other features, such as finish quality and corrosion resistance.

The main elements added to aluminum to obtain certain characteristics are:

	Element	Characteristics
Si	Silicon	Reduces the melting point while increasing fluidity in castings. Improves corrosion resistance.
Cu	Copper	Improves strength by promoting age hardening behavior. Decreases corrosion resistance.
Zn	Zinc	Allows heat treatment to improve strength.
Mg	Magnesium	Improves strength and corrosion resistance. Alloys will not age harden.
Mn	Manganese	Improves strain hardening without decreasing ductility. Improves corrosion resistance.
Fe	Iron	Common impurity. Sometimes added to improve strength.
Cr	Chromium	Improves toughness and controls grain structure growth.
Ti	Titanium	Helps refine grain structure. Added to weld wire.
Ni	Nickel	Improves hardness and strength.

There are many variations of aluminum alloys in use today. The Aluminum Association (AA) pegs the number at over 560 and growing. In art and architecture only, a few variations are considered for use and of these, only about a dozen are in common use.

At first it might appear that a designer can select and design an alloy for a particular application. You might have a chance if you have a significant quantity of metal needed for the project. Most mills will work with you to create the precise alloy and temper characteristics needed. However, if the quantity is not large, your choices are limited to the stocks produced for general use. This rarely poses an issue because there are several alloys designed for the art and architectural world. Most of these are available in many forms in lower quantities than a typical mill run.

ALLOY DESIGNATION SYSTEM

Each form of aluminum, wrought and cast, has specific designation systems for the alloys in use. Most people in the industry recognize the original four-digit designation system for the wrought forms and the three-digit system used for the cast forms. Variations on these by industry and by country exist. In this book we will use the system designated by the Unified Building System. This system has its basis in the original Aluminum Association designation.

Aluminum Association Four-Digit Code

1xxx series—The commercial pure alloying forms
2xxx series—Copper is the major alloying constituent in this series
3xxx series—Manganese is the major alloying constituent in this series
4xxx series—Silicon is the major alloying constituent in this series
5xxx series—Magnesium is the major alloying constituent in this series
6xxx series—Magnesium and silicon are the major alloying constituents in this series
7xxx series—Zinc is the main alloying constituent in this series
8xxx series—Other elements

The first number, 1–8, designates the particular alloy series. The alloy series or group number corresponds to the principal alloying element or elements added to aluminum. The second digit, if it is not zero, indicates specific modifications made to the base alloy of the series. For example, alloy 3105 would be the first modification to the 3005 alloy. The final two digits correspond to specific alloy types within the series.

This is the case for each series with the exception of the first series, the 1xxx series.

The 1xxx series, which is considered the commercial pure alloy forms of aluminum, must be at least 99% or greater in purity. The last two digits correspond to a purity level above 99%. For example, if the last two numbers are 5 and 0, then this would mean a purity level of 99.50%.

Unified Numbering System

There are several newer classifications that still relate back to the old four-digit series classifications. One is the Unified Numbering System (UNS) as recognized throughout North America. It will be used throughout this book. In the UNS system the prefix "A" is used for "aluminum" followed by a digit that designates wrought or cast alloys. The number 9 represents wrought alloys, while the number 0 represents cast aluminum alloys. There are a few others as well, but for all aluminum alloys they begin with the letter A (see Table 2.2).

A0xxxx cast alloys
A8xxxx wrought alloys clad with wrought aluminum alloys
A9xxxx wrought alloys

Alloy Designation System

TABLE 2.2 Wrought aluminum alloy designation.

UNS classification no. for wrought alloys	Original aluminum association's classifications for wrought alloys	Major alloying component
A91xxx	1xxx	Commercial pure
A92xxx	2xxx	Copper
A93xxx	3xxx	Manganese
A94xxx	4xxx	Silicon
A95xxx	5xxx	Magnesium
A96xxx	6xxx	Magnesium and silicon
A97xxx	7xxx	Zinc
A98xxx	8xxx	Other elements

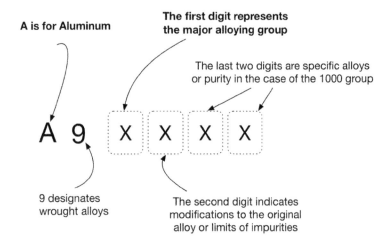

The last four digits are exactly as before with the Aluminum Associations system; only the prefix has been added to distinguish between wrought and castings.

Europe uses a similar system. Their system has the same last four digits and the corresponding alloying makeup that matches what is used in North America. The European system has the following prefixes to the four-digit alloy numbers.

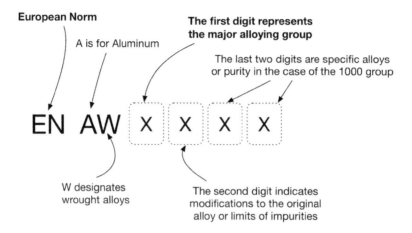

Tempers and Strengthening

The initial strength of aluminum is obtained by the addition of specific alloying elements. Further strength is induced by special treatments, called "Tempering" and referred to as "Temper" or "Heat Strengthening." This important strengthening process is performed at the mill or production facility in the final forms of the aluminum alloy.

Temper is a term used to describe the amount of stress or hardness imparted into aluminum. A distinctive temper designation is used to indicate what has happened to the aluminum after it is produced in order to impart a particular property of strengthening or hardening. The temper designation correlates to a particular alloy's overall strength. For aluminum, this is very important. Aluminum has the ability to undergo strength changes simply by resting at room temperature.

As Dr. Alfred Wilm discovered in 1906 with the invention of Duralum, certain strengthening behaviors can be modified with post processes to the metal. This is important for a lot of reasons. Fabrication processes such as stamping, deep drawing, forming, and welding induce changes in the metal at a granular level that will affect how it performs. With aluminum, temper is defined as the hardness and strength produced by mechanical means such as cold working or by thermal treatments that change the way the alloying elements interplay with the aluminum crystal lattice. Further, with aluminum a condition known as "aging" plays a role in the strengthening characteristics of the metal. The tempering process is critical when clarity and quality of finish are of concern.

Aluminum by itself, unalloyed, is soft. Cold working will improve the strength to a point, whereas alloying will improve the strength further. Aluminum, like all metals, is made up of crystalline atomic structures in the solid state. In order for the aluminum sheet or plate to be plastically deformed, this crystalline structure must be able to dislocate and slip in atomic layers. When things such as alloying elements are present, these can impede the slipping of the crystals and strengthen the metal. Cold working, otherwise known as strain hardening, create dislocations by altering the grain size and mixes them throughout the metal, breaking continuity in the crystal lattice and causing internal strain. This internal strain increases the strength of the metal because it acts as an internal counterforce to external imposed loads. Elevated temperature causes diffusion of the grains by making the atoms move around, which relieves some of the strain. The atoms move from highly strained regions to lower strained regions. As the temperature increases, new grains form and grow within the grains that are distorted by the cold working process. This will reduce the strength by reducing the internal strain.

In metallurgy, there is an additional behavior that develops, called "coherency." When the precipitate element atomic size is small, a continuous structure forms in the crystal lattice structure. This is called "coherent" and is stable. The crystal lattice is still distorted, but the mechanical properties are altered by the strain induced by the stress in the lattice (Figure 2.2).

When the particles reach a certain size within the crystal lattice, they can form their own crystal structure. The mechanical stresses disappear, and grain boundaries begin to develop. This is known as "noncoherent." With the formation of the grain boundaries, there is a decrease in corrosion resistance. This happens with the copper-bearing aluminum alloys. The copper molecule is larger and thus affects the aluminum crystal lattice in a noncoherent way, providing greater strength but at a loss of corrosion resistance.

There is a whole vocabulary associated with the tempers assigned to aluminum alloys.

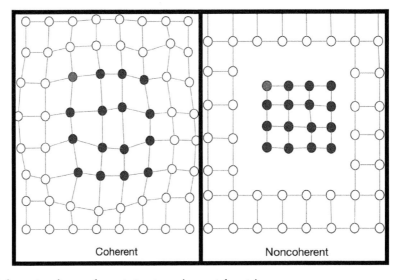

FIGURE 2.2 Coherent and noncoherent structures in crystal matrix.

TABLE 2.3 Tempers associated with aluminum alloys.

O	Fully annealed
F	As fabricated
T	Heat strengthened (thermal treatments)
H	Strain hardened (cold work treatments)
W	Solution heat treatment

There are four basic temper designations (Table 2.3). There is the -F category, which simply means "as fabricated." The metal is provided in an untested form. There is the -O, which is the fully annealed, dead soft temper. After these basic two, the tempers become much more defined.

For the majority of architectural alloys of aluminum, the T or H tempers are common. For art, the occasional O, fully annealed temper is used. It is suggested to not use the F temper in wrought product—it is more common in castings and some pipe. Not knowing the mechanical properties can lead to issues later on in fabrication and performance.

The W temper is also rarely used. It refers to solution heat treatment, but the treatment is considered unstable unless aging time is listed. Suppliers of aluminum wrought products will work diligently with you to achieve the required criteria and normally they process to specific H tempers or T tempers, depending on the alloy.

Within the alloy series there are certain, tempering constraints. A particular series can either be tempered by cold working or can be tempered by thermal treatments. Cold working is a term given to stressing the metal under the annealing temperature, which causes the metal to undergo plastic deformation as the grains are strained and stretched. Cold working is also called Strain Hardening and is designated by an H listed after the alloying number. The alloys that contain manganese, silicon, and magnesium fall within this grouping. These will not benefit from heat treatments.

Only alloys that fall into one of the following series can be strengthened by strain hardening:

A91xxx, A93xxx, and A95xxx

The other group can be strengthened by thermal treatments. The temper for these is designated by a T after the alloying number. The alloys that can be strengthened by thermal treatments contain the alloying elements copper, magnesium, zinc, and silicon. Heat treatment of these alloys enables the development of significant strength by subjecting them to thermal treatments as well as strain hardening. Many ferrous and nonferrous metals can have their strength increased by a combination of heat, rapid cooling, and cold working. Only the following series can be strengthened in this way.

A92xxx, A94xxx, A96xxx, and A97xxx

Strain hardening	Also known as *cold working* or work hardening. This increases hardness and strength and reduces ductility. The metal is passed at room temperature through a series of stressing operations.
Heat treatment	Also known as solution annealing or solution heat treatment. This makes the metal softer and more ductile. The metal is heated to a specific temperature and then cooled at a rapid rate.
Precipitation hardening	Also, known as *age hardening* or *aging*. There are two forms, Artificial Aging, in which after the metal has been annealed in the solution heat treatment process, it is reheated for a period of time below the solution heat treatment temperature, or Natural Aging, where strengthening develops with the passage of time after the Solution Heat Treatment process.

Heat Treatment of Aluminum

Specific alloys can be thermally treated to achieve significant strength increases by taking advantage of a phenomena where certain alloying elements are absorbed into the aluminum crystal lattice and are essentially trapped there. They distort the lattice structure which increases the strength of the aluminum by resisting dislocation movement of the aluminum crystals when strained. These alloys contain copper, magnesium, silicon, and zinc. It is these elements that act as solutes in what is referred to as a solid solution[1] of aluminum.

Alloys That Can Be Strengthened by Heat Treatment
A92xxx
A94xxx (certain alloys in this group)
A96xxx
A97xxx

Heat treatment, also known as "solution heat treatment," involves placing the aluminum form in a chamber and bringing the temperature up to levels where the added alloying elements go through solid solution with the aluminum. The temperature is around 548°C (1018°F). The temperature is maintained at these levels long enough for the atoms to become intermixed in the aluminum atom lattice. The aluminum then is quickly quenched in water to bring the temperature down rapidly and "freeze" the position of the alloying elements. The quenching takes only a few gseconds and it traps the atoms in the distorted aluminum lattice.

---

[1]We often think of a solution as liquid, but a solution can be a solid within a solid or a gas within a gas. In metallurgic terms, a solid solution is just that, a solid composed of two or more metal elements into a crystalline form that is homogeneous throughout.

Natural Aging and Artificial Aging

When the aluminum is stored at room temperature for a few days, it can gain hardness and stiffness. This is known as *Natural Aging*. As Dr. Alfred Wilm discovered, leaving certain alloys to lie at rest for a few days causes them to harden and increases their temper. This is known as natural aging and takes several days. This aging period allows the aluminum to reach its strength. This process can be sped up by performing a process known as "precipitation hardening," in which the metal is heated after quenching to a specific temperature and is held at that temperature for several hours. This process is known as *artificial aging*.

The letter T is used to define the particular heat treatment performed on the aluminum alloy. It is always followed by a number or series of numbers.

For example,

A96061-T6

The T denotes this aluminum alloy underwent thermal treatments to a particular level as defined by the number 6.

In architecture and art, for the wrought forms of sheet, plate, and extrusion there are several heat treatment tempers that are more commonly used than others. These represent the post processes performed on the wrought form. Of the heat strengthened alloys, the A96xxx are the alloys of architecture and art. The other heat-strengthened alloys are more often produced for the aerospace and automotive industries (Table 2.4).

Common Architectural Temper Designations
-T4
-T6

As manufacturing processes have expanded, more temper designations have been created. There are now many additions to the original listing. The first digit must be one of the above, then there are more digits added to define a particular attribute that is significantly altered from the original. For example, -T51, which represents cooling from an elevated temperature shaping process, then stress relieved by stretching, followed by artificial aging, is a further extension of the temper designations. There are more tempering designations listed in Appendix B. These are special, defined operations that modify the base temper defined by the first number value.

Strain Hardening (Nonheat Treatment of Alloys)

Strain hardening will occur when aluminum undergoes cold working. Cold working occurs when the aluminum undergoes a change in shape without the benefit of high temperature. For select

TABLE 2.4 The temper designations used with the various heat treatable alloys. The most common temper designations used in art and architecture are boldfaced.

Temper	Description
-T1	The metal is force cooled from the high temperature shaping process such as extruding or hot rolling. After this the metal is allowed to rest at room temperature for a period of time. This is natural aging where the aluminum strength characteristics stabilize. Not typically used. Corrosion resistance not as good as others and cold working mechanical properties have not been developed.
-T2	The metal is force cooled from the high temperature shaping process such as extruding or hot rolling. Following this the metal undergoes cold working before resting at room temperature to achieve natural aging to stabilize the strength characteristics. Not typically used because it lacks specific mechanical property considerations that other tempers impart.
-T3	The metal is solution heat treated where it is held at a particular temperature in a furnace, then hot worked followed by rapid quenching in water. Once cool the metal is further cold worked to develop strength. Then natural aging is used to stabilize the strength characteristics.
-T4	**The metal undergoes solution heat treatment, in which it is held at a particular temperature in a furnace. This is followed by natural aging to a stable condition.**
-T5	Cooled from a high temperature shaping process such as extruding and then artificially aged to enable precipitation hardening to occur. Essentially, in the artificial aging process, the temperature is held high for a designated period of time to allow the alloying constituents to enter into solid solution. The metal is cooled rapidly to "lock" the constituents into the crystal matrix.
-T6	**The metal undergoes solution heat treatment, in which it is held at a particular temperature in a furnace. This is followed by artificial aging to enable precipitation hardening to occur. In the artificial aging process, the temperature is held high for a designated period of time to enable the alloying constituents to enter into solid solution. This allows the metal to achieve maximum strength.**
-T7	Solution heat treatment, then over-aged in a furnace to beyond peak strength condition. Limited use of certain alloys to improve corrosion resistance.
-T8	Solution heat treated, cold worked to strain harden, artificially aged to achieve maximum strength.
-T9	Solution heat treated, artificially aged to achieve precipitation hardening, then cold worked to improve strength. Rarely used.
-Txx	A second digit is often added. If the number is a 1, 3–9, then a variation to the basic temper described in the first digit has been performed. Usually an alteration to the strength imparted to improve ductility. If the second number is a 2, it refers to heat treatment processes from any temper by the user of the product.
-Txxx	When a third digit is used, it represents a specific, additional process, usually cold working such as straightening or stress relief.
-Txxxx	When a fourth digit is used, special stress corrosion enhancement processes have been performed in combination with other treatments.

aluminum alloys, strength is added to various degrees from the processes of cold working alone. Stain hardening is a form of plastic deformation that changes the crystal lattice structure of the aluminum by inducing mechanical stress into the metal. It also changes or deforms the metal by stretching it while at the same time thinning the metal or shaping the metal. The strain can be measured and in the form of strain hardening, the strength of the aluminum can be predicted and measured.

For most sheet and plate material a desired strain hardening is achieved when the metal is passed between rollers that reduce its thickness. This is known as cold rolling. As this occurs, the grains of the metal are stretched and the elements within the aluminum alloy are locked into the crystalline matrix of the metal. As this occurs, the metal becomes further resistant to change. Sometimes the process of strain hardening goes beyond the desired hardness and strength and requires subsequent annealing to bring the metal to a particular hardness. Other strain-hardened tempers are given a heat treatment to stabilize the strength and hardness properties after cold working.

The process of cold rolling also provides the final finish from the mill. Various polished rolls and sometimes pattern rolls are used for the final passes to induce a finish into the surface while at the same time arriving at a strain hardened temper.

The most common method used to strain harden aluminum sheet is cold rolling the sheet through reducing rolls. Other methods involve pushing the metal through an orifice or die to reduce the size or subject the form to bending and straightening. Aluminum rods are strain hardened in this manner.

Alloys That Can Be Strengthened by Stain Hardening

A91xxx

A94xxx

A95xxx

There are four basic categories of strain hardening. Like the temper designation for the heat strengthened alloys, the strain hardening is described as "-H" followed by a series of numbers. For example:

A95005 -H32

The first digit represents one of the following:

- -H1   Strain hardened only
- -H2   Strain hardened and partially annealed
- -H3   Strain hardened and stabilized
- -H4   Strain hardened, followed by thermal heating from paint curing process

The second digit represents the degree of strain hardening that remains. The values are 1–9. The number 1 stands for fully annealed. The number 8 stands for fully hard. The numbers between 1 and 8 are degrees between fully annealed and fully hard. The number 9 stands for a strength imparted that is at least 2 ksi greater than fully hard, also known as "extra hard" temper.

For each alloy there is a maximum hardness, commonly referred to as "full hard." More force is needed to bend alloys that are full hard temper. The industry further uses descriptions of degrees of hardness. Common are "quarter hard," "half hard," and "three-quarter hard," as well as fully annealed (soft), "full hard" and "extra hard." The values depend on the alloy, but the numbering closely follows what is referred to as cold work hardness and described as a proportion to the hardness achieved when the alloy is in the fully hard state.

1: Fully annealed
2: Quarter hard temper
4: Half hard temper
6: Three-quarter hard temper
8: Fully hard temper

The odd digits represent intermediate levels with the number 9 being extra-hard temper.

The Aluminum Association allows a third digit for strain hardened parts that represents a variation. The variation must be close to the temper defined for the alloy on the second digit. This third digit is a customized characteristic that is optional and defined by the manufacturer (Table 2.5).

Each of these temper values correspond to mechanical properties of the particular alloy. They also adjust the grain makeup of aluminum and thus the surface of the aluminum. The specific temper should be discussed with the supplier of the wrought product, sheet, plate, or extrusion to understand precisely what temper is being provided and what this means to the fabrication and mechanical properties of the aluminum.

A fabrication facility may want to adjust the temper because specific forming or shaping operations will be undertaken on the metal. As the metal is further cold worked at the fabrication facility, the hardness increases incrementally. In certain instances, it may be desirable to start with a lower temper. If the end product is going to be shaped significantly, say by hammering or spinning, you may wish to order a lower temper and allow the cold work performed on the metal to work harden the aluminum to a higher temper.

The initial strength of aluminum is obtained by the addition of specific alloying elements. Further strength is induced by special treatments, called "Tempering" and referred to as "Temper" or "Heat Strengthening." *Temper* is a term used to describe the amount of stress or hardness imparted into aluminum.

TABLE 2.5 Temper designations used on nonheat treatable aluminum alloys.

Temper	Number	Description
H	12	Cold worked to a quarter hard temper
	14	Cold worked to a half hard temper
	16	Cold worked to three-quarter hard temper
	18	Cold worked to full hard temper
	22	Cold worked then partially annealed to quarter hard
	24	Cold worked then partially annealed to half hard
	26	Cold worked then partially annealed to three-quarter hard
	28	Cold worked then partially annealed to full hard
	31	Cold worked then stabilized to a reduced, full annealed state
	32	Cold worked then stabilized to reduced quarter hard state
	34	Cold worked then stabilized to reduce half hard state
	36	Cold worked then stabilized to reduce to a three-quarter hard state
	38	Cold worked then stabilized to full hard state
	X11	Cold worked to a slight degree over fully annealed, but less than HX1 classification
e.g.	112	Strain hardened from working at an elevated temperature
e.g.	311	Cold worked to a degree less than H31
e.g.	321	Cold worked to a degree less than H32
e.g.	323–343	Products that underwent cold working to reduce the occurrence of stress-corrosion cracking

Wrought Alloys Used in Art and Architecture

There are several alloy types used today in art and architecture. These are listed by the main alloying groups and common temper designations where applicable are described. These do not represent the entirety of the family of aluminum alloys but only those that find their way into art and architecture (Table 2.6).

The majority of the alloys used in art in architecture fall into the A93xxx, A95xxx, and A96xxx alloy groups. These have good clarity in appearance combined with superior corrosion resistance. These alloys can be anodized to protect their surface and will take a polish to enhance their

TABLE 2.6   List of the alloys discussed in this chapter.

Wrought alloy	Major alloying element	Characteristic
A91100	Commercial pure	Low strength, good ductility, and corrosion resistance. Bright appearance and good clarity.
A92014	Copper	High strength, poor corrosion resistance. Usually clad with another alloy with better corrosion resistance.
A92017	Copper	Good strength, poor corrosion resistance. Fasteners are made from this alloy.
A92024	Copper	Good strength, poor corrosion resistance. This alloy was once called Duralum in World War II
A93003	Manganese	Multipurpose architectural alloy. Good corrosion resistance. Good ductility.
A93004	Manganese	Multipurpose architectural alloy. Good corrosion resistance. Good ductility.
A93105	Manganese	Good strength. Good corrosion resistance. Made up of mostly recycled aluminum. Poor anodized appearance.
A94043	Silicon	This alloy is used for welding rod and welding wire.
A95005	Magnesium	Major architectural alloy. Good strength, good corrosion resistance, and good surface clarity. Anodizes well.
A95050	Magnesium	Good strength and good ductility.
A95052	Magnesium	Higher strength. Good clarity and will anodize well.
A95056	Magnesium	High strength alloy used for fasteners.
A95083	Magnesium	Good corrosion resistance. Considered a marine alloy.
A95086	Magnesium	Good corrosion resistance. Considered a marine alloy.
A95154	Magnesium	Good corrosion resistance. Good ductility.
A95754	Magnesium	Good corrosion resistance. Considered a marine alloy.
A96005	Magnesium and silicon	Good corrosion resistance. Alloy used to produce extrusions.
A96053	Magnesium and silicon	Good corrosion resistance. This alloy is used to make aluminum wire.
A96060	Magnesium and silicon	Excellent extrusion alloy. Complex shapes and good finish quality can be achieved.
A96061	Magnesium and silicon	Major architectural alloy. Good corrosion resistance. Good surface clarity and will anodize well.
A96063	Magnesium and silicon	Major architectural alloy. Used for extrusions. Good color and good corrosion resistance.
A96151	Magnesium and silicon	High strength alloy. Lower ductility. Used for forging.
A96463	Magnesium and silicon	Good strength. Good finish clarity. Used for architectural trim and extrusions.
A97075	Zinc	Aerospace alloy. Very high strength.

appearance. These alloy groups have good ductility and adequate strength. The A96xxx group can be heat treated while the A93xxx and A95xxx can be strain hardened. Each of these alloy groups have superior corrosion resistance with the A95xxx, offering good marine environment performance.

A91xxx Series

This series represents the highly pure form of aluminum. The aluminum must be at least 99% pure. Impurities include silicon and iron in trace amounts. This group is common in the electrical industry because of its purity. The higher the purity percentage, the better the electrical and thermal conductivity. These alloys are highly corrosion resistant, again due to the purity, however, they are incapable of developing good strength. Strain hardening will increase the alloys' strength to a point. They cannot be heat treated.

These alloys are malleable and soft. Their lack of surface hardness makes them prone to being scratched. They can be anodized. Clear anodized surfaces have excellent clarity and consistency. Bright dipped anodizing or electropolishing provides a highly reflective surface. When anodized, the surface quality is superior to other aluminum alloys.

Often these alloys are clad over other, less corrosion resistant alloys such as the A92xxx and A97xxx alloys. The purity of the A91xxx alloys affords excellent corrosion protection but the thin layer of clad surface is soft and prone to being scratched.

In art and architecture, the alloys that find occasional use are listed together because they are distinguished by the levels of purity: 99.30% pure, 99.50% pure, 99.80% pure, and 99.90% pure are several common alloys. The A91100 alloy is 99% pure and contains 0.12% copper to give it strength. It is the strongest of the A91xxx series alloys.

Alloy A91030	99.30% aluminum
Alloy A91050	99.50% aluminum
Alloy A91080	99.80% aluminum
Alloy A91090	99.90% aluminum
Alloy A91100	99.00% aluminum

Each of these alloys are high purity forms with A91100 being the strongest of the series. The A91xxx alloys cannot be strengthened by heat treatment processes, and limited benefits are gained through strain hardening by cold working. These alloys have superior luster and are used where significant forming is required. They are also used in solar reflectors and light reflectors because of their ability to be chemically polished and have superior reflectance to other alloys and metals due to their high purity. They are often clad over the surface of other stronger, but less corrosion-resistant, alloys. The cladding process, where one alloy of aluminum is thinly coated over another is referred to as "alclad."

The A91xxx series is weldable. The filler rod should be the same alloy.

ALUMINUM ALLOY A91100

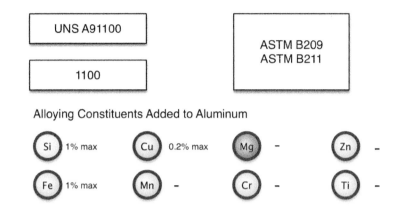

Alloy	Tensile strength		Yield strength		Elongation % (2 in.–50 mm)	Rockwell Hardness scale
A91100-H14	18 ksi	125 MPa	17 ksi	117 MPa	9	32
A91100-H18	24 ksi	165 MPa	22 ksi	152 MPa	5	44

These highly pure alloys of aluminum are very ductile and soft. There is not a great improvement in strength when cold worked from half hard to a full hard temper.

Available Forms of Alloy A91100
Sheet
Plate
Foil
Extrusion
Bar
Tube
Clad (coating)

Alloy A91100 is used in diverse applications. It is a high purity alloy form with excellent ductility. This alloy lacks strength but has excellent clarity and superior corrosion resistance. Strength can be increased by cold working. It cannot be strengthened by heat treatment processes.

A92xxx Series

The alloys in this series have higher levels of copper. Copper is common as a trace element in many of the alloy forms. Copper has good solubility in the aluminum crystal makeup. Copper promotes the age hardening behavior of the alloys. This is the only series where some alloy forms can be thermally treated to alter mechanical properties, while others require strain hardening. The A92xxx has good strength. They were the first to be heat strengthened and have been in use in structural applications from the beginning of the industrial production of aluminum. They are the most susceptible to corrosion and many applications needing strength are often clad with higher purity aluminum alloys to provide the corrosion protection they lack. They can be anodized, but the appearance is variable and inconsistent as the etching process leaves copper smut on the surface.

These alloys can be clad in the A91xxx alloys to improve corrosion resistance.

These alloys can be welded using A92xxx filler. Some of the alloys are not weldable by arc welding processes.

ALUMINUM ALLOY A92014

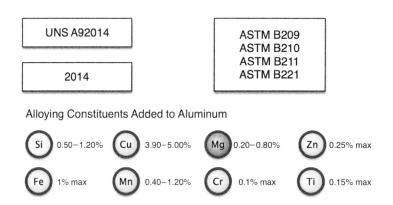

Alloy	Tensile strength		Yield strength		Elongation % (2 in.–50 mm)	Brinell hardness number
A92014-T6	70 ksi	483 MPa	60 ksi	414 MPa	13	135

Available Forms of Alloy A92014
Sheet
Strip
Plate
Extrusion
Bar

Alloy A92014 is very high in strength. It is not commonly used in architecture or art, it is one of the strongest alloys of aluminum available. Considered an aerospace alloy, it is usually clad with a higher purity alloy, such as A91100, to improve corrosion resistance.

Alloy Designation System

ALUMINUM ALLOY A92017

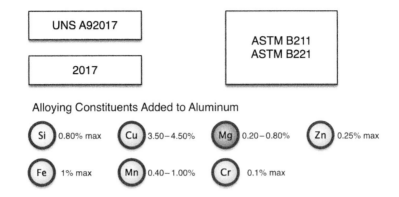

Alloy	Tensile strength		Yield strength		Elongation % (2 in.–50 mm)	Brinell hardness number
A92017-O	26 ksi	179 MPa	10 ksi	69 MPa	22	45
A92017-T4	62 ksi	427 MPa	40 ksi	276 MPa	22	105

Available Forms of Alloy A92017
Fasteners
Bar

This alloy is used as rivets and stressed skins in the aerospace industry. Strength is improved by heat treatment processes. This is not a common architectural or sculptural alloy.

ALUMINUM ALLOY A92024

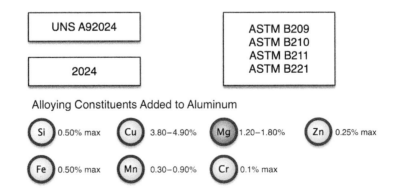

Chapter 2 Aluminum Alloys

Alloy	Tensile strength		Yield strength		Elongation % (2 in.–50 mm)	Brinell hardness number
A92024-T4	68 ksi	469 MPa	47 ksi	324 MPa	19	120

Available Forms of Alloy A92024

Sheet
Plate
Extrusion
Bar

This alloy is stronger than A92017. Alloy A92024 is commonly extruded. It is considered an aerospace alloy with few examples of use in art or architecture, and was the workhorse of World War II. At that time it was known as alloy 24S. It was also very similar in makeup to the early alloy created by Dr. Welm and named "Duralumin." This alloy was used to produce nearly 300 000 planes for the war effort. The Ford Tri-Motor airplane built in the mid 1920s was wrapped in corrugated Duralumin with an alclad coating of high purity aluminum for corrosion protection (Figure 2.3).

FIGURE 2.3 Ford Tri-Motor plane and an image of the corrugated wing cladding.

A93xxx Series

These are manganese-bearing alloys. Manganese has low solubility in aluminum but will provide good corrosion resistance. The most common form of these alloys is thin sheet. The A93xxx is the basis of the aluminum can industry. Plate, extrusion, and tube forms are also available. The A93xxx alloys are some of the most common of the aluminum alloys used in the building industry and appliance industry. The manganese-bearing alloys have excellent formability and adequate strength characteristics. Sheets made from these alloys are used as flashing, roll formed panels and other light industry fabrications that need good ductility coupled with good corrosion resistance. They have moderate strength and are good for high temperature applications. Most aluminum cooking utensils are made from alloys in this series, as well as the ubiquitous aluminum can. These alloys can be welded with A91xxx, A94xxx, and A95xxx filler alloys. They are strengthened by cold working.

ALUMINUM ALLOY A93003

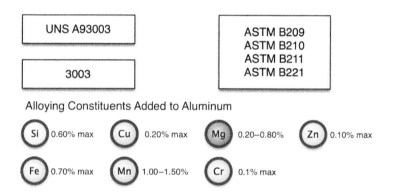

Alloy	Tensile strength		Yield strength		Elongation % (2 in.–50 mm)	Brinell hardness number
A93003-O	16 ksi	110 MPa	6 ksi	40 MPa	30	28
A93003-H14	22 ksi	152 MPa	21 ksi	145 MPa	8	40

The H14 temper is just one of several available to this alloy. H14 stands for strain hardening by cold working to a point where half hard temper is achieved. Note the increase in yield strength that is achieved over the O-temper. The O-temper would only be used if significant cold work is intended to be performed such as deep drawing, hammering, or spinning. These operations work harden the metal and increase the temper from intense cold shaping operations.

Available Forms of Alloy A93003
Sheet
Plate
Foil
Extrusion
Bar
Tube
Clad (coating)

Alloy A93003 is a general-purpose sheet metal alloy. It has good corrosion resistant characteristics and is used as gutters, siding, roofing, and other utility forms of aluminum. It is often coated with paint in coil coating processes. It can be welded and strengthened by cold working.

ALUMINUM ALLOY A93004

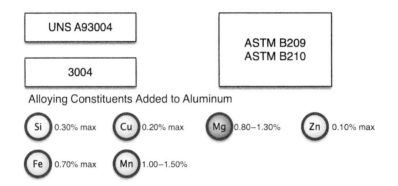

Alloy	Tensile strength		Yield strength		Elongation % (2 in.–50 mm)	Brinell hardness number
A93004-O	26 ksi	179 MPa	10 ksi	69 MPa	20	45
A93004-H34	35 ksi	241 MPa	29 ksi	200 MPa	9	63

The H34 temper is achieved by cold working the sheet material in cold rolling operations, followed by a room temperature aging treatment to stabilize the strength obtained. In this case, half hard temper was achieved in the process of cold rolling. The yield strength nearly triples the O-temper yield strength.

Available Forms of Alloy A93004
Sheet
Plate
Tube

This alloy and its near twin, A93104, is produced primarily for the aluminum can industry with a temper designation of H19. It is also used in general sheet metal as flashings and other basic industrial metal forms such as metal siding and roofing.

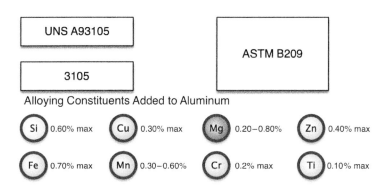

ALUMINUM ALLOY A93105

Alloy	Tensile strength		Yield strength		Elongation % (2 in.–50 mm)	Brinell hardness number
A93105-O	17 ksi	117 MPa	8 ksi	55 MPa	24	—
A93105-H14	25 ksi	172 MPa	22 ksi	152 MPa	5	—

Cold working has a profound effect on the strength of imparted to alloy A93105, which achieves nearly three times the yield strength over the O-temper form.

Available Forms of Alloy A93105

Sheet

This is a general purpose aluminum alloy used in the building industry. It has better strength than the A93003 aluminum alloy. Alloy A93105 is made from at least 95% recycled aluminum. It does not anodize well but it can be painted. Of all the alloys of aluminum, A93105 can have a range of alloying constituents in various combinations and amounts. This accounts for the variations in alloying constituents coming from the recycled material.

A94xxx Series

These alloys are not commonly used in architect as wrought alloys. These silicon-bearing alloys are more common in casting due to the lower melting point enabled by the silicon. They are also used as welding wire and rod. Some of these alloys can be heat treated, while others can gain strength only by cold working. It is the only A94xxx series alloy listed here. Alloy A94043 is a commonly used welding wire for architectural assemblies. The added silicon lowers the melting point slightly and this allows the filler to melt at a lower temperature than the base metal.

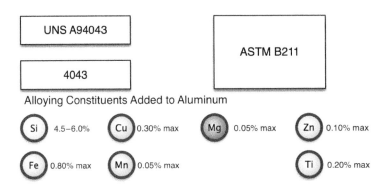

ALUMINUM ALLOY A94043

Alloy	Tensile strength		Yield strength		Elongation % (2 in.–50 mm)	Brinell hardness number
A94043-O	21 ksi	145 MPa	10 ksi	70 MPa	22	—

Available Forms of Alloy A94043

Welding rod or welding wire

This alloy is commonly used as welding wire or rod. This high silicon filler metal alloy is used to weld the heat treatable alloys of A96xxx series. It has good corrosion resistance, but when anodized will appear darker than the surrounding metal (Figure 2.4).

A95xxx Series

The A95xxx series is the category where most wrought sources for art and architectural aluminum are obtained. These alloys are high in aluminum purity with magnesium as the major element introduced. Magnesium is soluble in aluminum and provides an alloy with good strength, ductility, and durability. These alloys are used in many applications where excellent corrosion resistance is demanded along with excellent surface quality. Some of these alloys, A95083, A95086, and A95754 are considered as marine-grade alloys and are suitable for coastal environments.

The A95xxx alloys can be anodized, and several are the preferred alloy for color anodizing. They can be welded, polished, formed, and shaped. Typical forms are sheet and plate. They are nonheat treatable and gain strength through cold working. The A95xxx alloys listed here will not age harden.

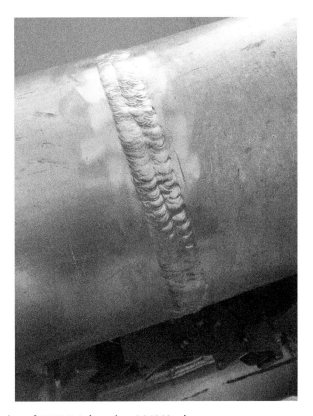

FIGURE 2.4 Welded section of A96061 tube using A94043 wire.

ALUMINUM ALLOY A95005

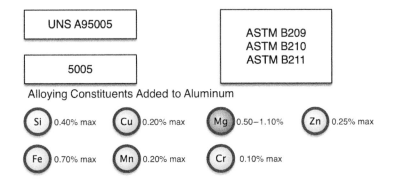

FIGURE 2.5 A95005-H34 anodized aluminum on the Grace Farms, designed by SANAA.

Alloy	Tensile strength		Yield strength		Elongation % (2 in.–50 mm)	Brinell hardness number
A95005-H16	26 ksi	179 MPa	25 ksi	172 MPa	5	—
A95005-H32	20 ksi	139 MPa	17 ksi	117 MPa	11	36
A95005-H34	23 ksi	159 MPa	20 ksi	139 MPa	8	41
A95005-H36	26 ksi	179 MPa	24 ksi	165 MPa	6	46

These are common tempers for the A95005 alloys. Often A95005 sheet is painted and will be designated as A95005-H42 to accommodate changes in temper from oven temperatures used to cure the paint.

FIGURE 2.6 A95005-H34 gold anodized sheet used on the Haas Motor Museum.
Source: Designed by Stocker Hoesterey Montenegro Architects.

Available Forms of Alloy A95005

Sheet

Plate

Bar

Alloy A95005 is one of the major architectural alloys. It has good strength and good corrosion resistance. It can be polished and buffed to produce a bright reflective surface. Major forms are sheet and plate. This alloy A95005 has excellent clarity and will anodize better and clearer than the A93xxx alloys. For color anodized sheet, this alloy is preferred because of the clarity of the oxide that develops and the good pore formation on the surface (Figure 2.5 and Figure 2.6).

Certain producers of this alloy in sheet and plate will further distinguish the finish for anodizing as AQ (Anodized Quality). This certifies that the surface will be free of streaking and other anomalies that may manifest on the surface after anodizing (Figure 2.7).

FIGURE 2.7   Engineered frames using A95005-H32.
Source: Metalabs office space.

ALUMINUM ALLOY A95050

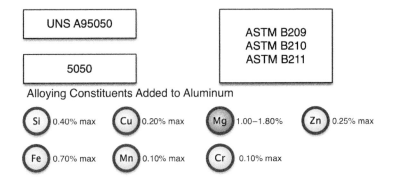

Alloy	Tensile strength		Yield strength		Elongation % (2 in.–50 mm)	Brinell hardness number
A95050-H32	25 ksi	172 MPa	21 ksi	145 MPa	9	46

Available Forms of Alloy A95050
Sheet
Plate
Bar
Tube

Alloy A95050 is a flat rolled product. It has very good workability. Architectural uses are for trim and hardware. Alloy A95050 is similar to A95005 but does not produce an anodized finish with as good a clarity. The biggest difference is the increase in allowable magnesium.

ALUMINUM ALLOY A95052

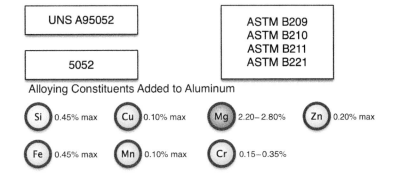

Alloy	Tensile strength		Yield strength		Elongation % (2 in.–50 mm)	Brinell hardness number
A95052-H32	33 ksi	228 MPa	28 ksi	193 MPa	12	60
A95052-H36	40 ksi	276 MPa	35 ksi	241 MPa	8	73

Available Forms of Alloy A95052
Sheet
Plate
Foil
Bar
Tube

This alloy, A95052, can achieve higher strength but with good ductility.

This alloy can anodize well, like the other A95xxx alloys. The higher strength of this alloy achieved from cold working makes it an ideal choice for many architectural applications subjected to stress (Figure 2.8).

ALUMINUM ALLOY A95056

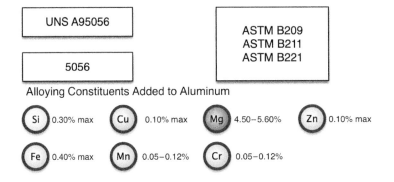

Alloy	Tensile strength		Yield strength		Elongation % (2 in.–50 mm)	Brinell hardness number
A95056-H18	63 ksi	434 MPa	59 ksi	407 MPa	10	105

Available Forms of Alloy A95056
Foil
Bar
Wire

FIGURE 2.8   Black anodized aluminum alloy A95052-H32, by Jan Hendrix.

Alloy A95056 is not a common architectural alloy. It is listed here because of its higher tensile strength, the highest among the A95xxx alloys. It is used for fasteners and wire but is also available in other wrought forms.

ALUMINUM ALLOY A95083

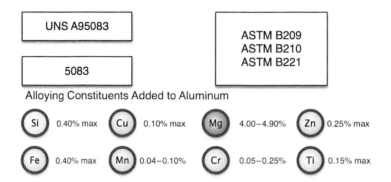

Alloy	Tensile strength		Yield strength		Elongation % (2 in.–50 mm)	Brinell hardness number
A95083-H112	44 ksi	305 MPa	28 ksi	195 MPa	16	—
A95083-H321	46 ksi	317 MPa	33 ksi	230 MPa	16	—

Available Forms of Alloy A95083
Sheet
Plate
Extrusion
Tube

The alloy, A95083, along with A95086 is considered an excellent marine-grade alloy of aluminum. Alloy A95083 has good resistance to chlorides. It has good strength and can be anodized. It is used as sheathing for ships and boats. Alloy A95083 has good weldability and strength.

ALUMINUM ALLOY A95086

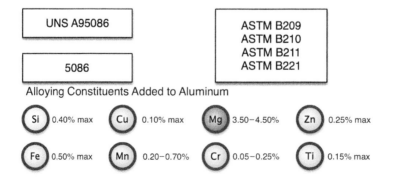

Alloy	Tensile strength		Yield strength		Elongation % (2 in.–50 mm)	Brinell hardness number
A95086-H32	42 ksi	290 MPa	30 ksi	207 MPa	12	—
A95086-H34	47 ksi	324 MPa	37 ksi	255 MPa	10	120

Available Forms of Alloy A95086
Sheet
Plate
Extrusion
Tube

Alloy A95086 is an excellent marine-grade alloy of aluminum. It can be welded without a great loss of strength. It is used for the hulls of small boats and ships. It has higher strength levels than A95050 and slightly higher strength than A95083 in the H34 temper.

ALUMINUM ALLOY A95154

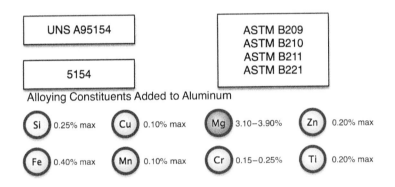

Alloy	Tensile strength		Yield strength		Elongation % (2 in.–50 mm)	Brinell hardness number
A95154-H36	45 ksi	310 MPa	36 ksi	248 MPa	12	78

Available Forms of Alloy A95154
Sheet
Plate
Extrusion
Bar
Tube

Alloy A95154 is a marine grade alloy of aluminum. It possesses good corrosion resistance and good strength. It is not as common as the A95083 or A95086 alloys.

ALUMINUM ALLOY A95754

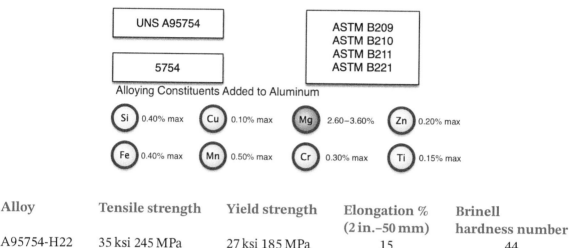

Alloy	Tensile strength	Yield strength	Elongation % (2 in.–50 mm)	Brinell hardness number
A95754-H22	35 ksi 245 MPa	27 ksi 185 MPa	15	44

Available Forms of Alloy A95754
Sheet
Plate
Extrusion
Bar
Tube

Alloy A95754 is a marine grade alloy of aluminum. It possesses good corrosion resistance and good strength in seawater and polluted exposures. Alloy A95754 is used on ships and as aluminum tread plate. It is also one of the alloys that is continuous cast to produce sheet and coil material.

A96xxx Series

The A96xxx series of wrought aluminum alloys is another common alloy group used in art and architecture. Some of the alloys are very similar to those of the A95xxx series. They both have magnesium as a major alloying element, and some of the A95xxx series can possess silicon into ranges that encroach into the A96xxx series levels. The major difference is the tempering process. The A96xxx alloys can be heat treated to develop strength, whereas the A95xxx alloys gain their strength by strain hardening. A96xxx alloys make up the majority of aluminum alloy extrusions used in art and architecture. These forms are not easily subjected to cold working process and so heat treatment becomes the preferred method of obtaining strength.

These alloys can be welded with A94xxx or A95xxx filler material.

ALUMINUM ALLOY A96005

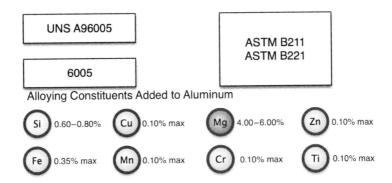

Alloy	Tensile strength		Yield strength		Elongation % (2 in.–50 mm)	Brinell hardness
A96005-T4	29 ksi	200 MPa	14 ksi	100 MPa	8	90
A96005-T5	42 ksi	290 MPa	39 ksi	270 MPa	9.5	90
A96005-T6	44 ksi	300 MPa	30 ksi	250 MPa	11	95

Available Forms of Alloy A96005
Extrusion
Tube

The A96005 alloy had good extruding qualities. It can be welded. Filler wire type A94043 wire is used. The typical wrought form of this alloy is extruded tube.

ALUMINUM ALLOY A96053

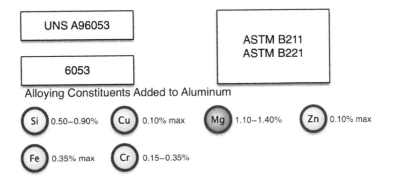

Chapter 2   Aluminum Alloys

Alloy	Tensile strength		Yield strength		Elongation % (2 in.–50 mm)	Brinell hardness number
A96053-T6	37 ksi	255 MPa	32 ksi	220 MPa	35	—

Available Forms of Alloy A96053

Wire

This alloy is typically in the form of wire or rod material. It is sometimes referred to as "aircraft wire" and used to hang architectural or art features or incorporated into woven aluminum wire products.

ALUMINUM ALLOY A96060

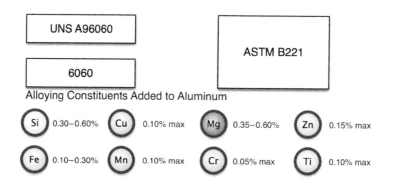

Alloy	Tensile strength		Yield strength		Elongation % (2 in.–50 mm)	Brinell hardness number
A96060-T4	20 ksi	140 MPa	10 ksi	71 MPa	16	
A96060-T5	23 ksi	160 MPa	17 ksi	110 MPa	9	
A96060-T6	32 ksi	220 MPa	24 ksi	170 MPa	11	—

Available Forms of Alloy 6060

Extrusion

Alloy A96060 is very similar to alloy A96063. The major difference is a lower magnesium level. This alloy extrudes very well. Very complex profile shapes can be created with alloy A96060. The finishing is also of good quality. This alloy is used to extrude profiles in the architectural facade industry.

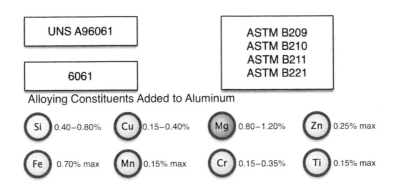

ALUMINUM ALLOY A96061

Alloy	Tensile strength		Yield strength		Elongation % (2 in.–50 mm)	Brinell hardness number
A96061-O	18 ksi	125 MPa	8 ksi	55 MPa	25	30
A96061-T4	35 ksi	240 MPa	21 ksi	145 MPa	22	65
A96061-T6	45 ksi	310 MPa	40 ksi	275 MPa	12	95

Available Forms of Alloy A96061

Sheet

Plate

Extrusion

Bar

Tube

This alloy exhibits excellent corrosion resistance in most environments. The surface gives exceptional anodizing quality. The quality of the surface in plate forms is superior to that of the A95005 series (Figure 2.9).

FIGURE 2.9 Gate made from 13 mm thick A96061-T651 and polished, by Jan Hendrix.

ALUMINUM ALLOY A96063

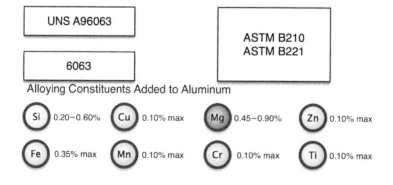

Alloy	Tensile strength		Yield strength		Elongation % (2 in.–50 mm)	Brinell hardness number
A96063-O	13 ksi	90 MPa	7 ksi	48 MPa	—	25
A96063-T1	22 ksi	150 MPa	13 ksi	90 MPa	20	42
A96063-T4	25 ksi	170 MPa	13 ksi	22 MPa	22	—
A96063-T5	27 ksi	185 MPa	21 ksi	145 MPa	12	60
A96063-T6	35 ksi	240 MPa	31 ksi	215 MPa	12	73

Available Forms of Alloy A96063
Extrusion
Tube

Alloy A96063 is the most common extruded alloy of aluminum. This alloy exhibits good corrosion resistance in most environments. The surface of this alloy gives good anodizing clarity. Strength and hardness levels can be improved significantly from heat treatment processes. It is an alloy commonly used on curtainwall extrusions and storefronts. This alloy can be welded, but expect a 30% decrease in strength. As a heat treatable alloy, some amount of strength can be restored by subsequent heat treatments (Figure 2.10).

FIGURE 2.10 Aluminum bench. Made from A96063-T4 and 96061 Plate. Cast aluminum legs. Designed by Jonathan Olivares.

ALUMINUM ALLOY A96151

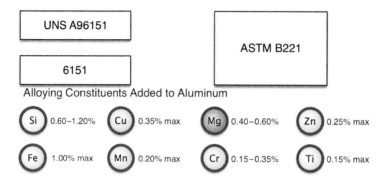

Similar to A96063. Extruded form used for forging.

Alloy	Tensile strength		Yield strength		Elongation % (2 in.–50 mm)	Brinell hardness number
A96151-T6	32 ksi	220 MPa	28 ksi	195 MPa	15	71

Available Forms of Alloy A96151

Extrusion used for forging parts.

Alloy A96151 has good strength but sacrifices ductility. It is used in forgings. The automotive industry uses this alloy in many of its forged assemblies. Not in common architectural use.

ALUMINUM ALLOY A96463

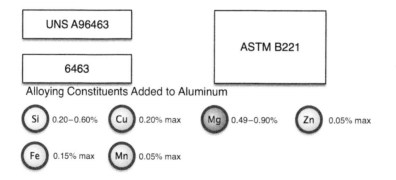

Alloy	Tensile strength		Yield strength		Elongation % (2 in.–50 mm)	Brinell hardness number
A96463-T1	22 ksi	150 MPa	10 ksi	90 MPa	20	42
A96463-T5	26 ksi	180 MPa	20 ksi	130 MPa	10	60
A96463-T6	34 ksi	230 MPa	29 ksi	200 MPa	11	74

Available Forms of Alloy A96463

Extrusion

Alloy A96463 is very similar to alloy A96063. Extensively used in architectural and ornamental trim.

A97xxx Series

These alloys are not commonly used in architecture or art; instead, they are considered suitable for the aerospace industry. These alloys possess very high strength, but corrosion resistance is poor. They contain zinc as the major alloying constituent. Only one alloy is listed, A97075, which is one of the more common aerospace alloys with extremely high strength when tempered. The yield strength of the T6 temper exceeds some steels. The alloy can be welded, and A95xxx filler is commonly used.

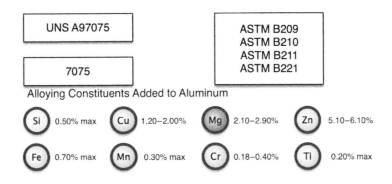

ALUMINUM ALLOY A97075

UNS A97075

ASTM B209
ASTM B210
ASTM B211
ASTM B221

7075

Alloying Constituents Added to Aluminum

Si 0.50% max, Cu 1.20–2.00%, Mg 2.10–2.90%, Zn 5.10–6.10%
Fe 0.70% max, Mn 0.30% max, Cr 0.18–0.40%, Ti 0.20% max

Alloy	Tensile strength		Yield strength		Elongation % (2 in.–50 mm)	Brinell hardness number
A97075-0	38 ksi	230 MPa	15 ksi	105 MPa	17	60
A97075-T6	83 ksi	570 MPa	73 ksi	505 MPa	11	150

Available Forms of Alloy A97075
Sheet
Plate
Extrusion
Bar
Tube

The alloy A97075 is an aerospace alloy first secretly developed for aircraft iby the Imperial Japanese Navy during World War II. Initially this extreme strength alloy was used on propellers for aircraft. It later became the structure for the Boeing B29 Super Fortress used during the later periods of World War II. It was once called ESD (extra super Duralumin) aluminum because of its high strength. This very strong and hard alloy of aluminum can be given a highly polished surface. It is used where high strength is needed.

Cast Alloys in Art and Architecture

Aluminum entered the world of art and architecture in the early 1900s as a cast metal. The first projects of the new era of manufacturing the metal centered on sand casting aluminum into artistic architectural embellishments. The cast steeple of the Smithfield Church in 1927 is one of the first examples in the United States. Located in the home of early aluminum production, Pittsburgh, Pennsylvania, this massive cast assembly is anchored to a steel inner structure. It still performs its decorative function nearly 100 years later.

Cast aluminum is a common form of the metal in many consumer and industrial applications. In art and architecture there is a resurgence in decorative cast aluminum panels and decorative parts.

Aluminum Alloy Series Designations: Cast Form

A designer considering cast aluminum needs to weigh several factors and how they influence the end quality and cost, for example, how the casting is to perform, the quantity of castings of the same or similar shape, what the final finish will be, and the size and shape of the piece. Additionally, if the aluminum casting is to have specific engineered mechanical properties, then the alloy and heat treatment requirements are important considerations.

The casting technique will determine the surface quality and level of tolerances achievable. For instance, if the final finish of the cast aluminum surface and tight tolerances of the final shape are critical, then consider near net-shape casting. Near net-shape casting increases the mold cost but reduces the final finishing costs. Large parts may be better suited to sand casting but the machining or finishing of the final surface may add cost and there may be a trade-off in tolerances.

One of the biggest drawbacks to simple aluminum casting is porosity in the cast part. Porosity is caused by hydrogen bubbles in the molten pour. Hydrogen is readily absorbed into molten aluminum and comes from moisture in the air. The hydrogen gas forms bubbles or pores near the

Alloy Designation System 79

FIGURE 2.11 Example of a cast aluminum façade.

surface as the metal solidifies. The bubbles that form near the surface of the mold can lead to a very poor cast surface finish. Aluminum oxide may form within the casting around small inclusions in the metal. This can lead to a weak casting and produce voids and pockets.

Specifying the Alloy

The systems used to specify aluminum castings can be very confusing. There is no widely accepted classification system for designating a particular alloy. The Aluminum Association (AA) and the American National Standards Institute (ANSI) have tried to correct this, as has the Unified Numbering System for Metals and Alloys (UNS), but there is still a jargon that has persisted in the nonferrous casting industry. Both the AA designation as well as the UNS will be shown. Specific alloys that are common to the world of architecture and art will be described.

For the cast forms of aluminum there is a similar series designation system for the series. Casting uses a three-digit number system followed by a decimal point and single digit. The first digit, similar to the wrought alloys, designates the major alloying elements.

1xx.x series—The commercial pure alloying forms
2xx.x series—Copper is the major alloying constituent in this series
3xx.x series—Silicon plus copper and/or magnesium are alloying constituents included in this series
4xx.x series—Silicon is the major alloying constituent in this series
5xx.x series—Magnesium is the major alloying constituent in this series
6xx.x series—Not currently in use
7xx.x series—Zinc is the main alloying constituent in this series
8xx.x series—Tin is the main alloying constituent in this series
9xx.x series—Other elements

The second and third digits designate a specific alloy in the series with the exception of the 1xx.x series. In this case, the second and third digits correspond to the level of purity. 100.1 stands for an aluminum ingot with 99.00% aluminum. Alloy 150.1 would be an ingot with an aluminum content of 99.50%.

The last value is either a 0, 1, or 2. A 0 represents it as a casting. The number 1 represents the alloy as an ingot to be used to make the casting, while a 2 represents this alloy as an ingot with tighter, more specific, chemical limits. For the final casting used in an architectural or artistic assembly, only the xxx.0 classification should apply.

The xxx.1 and xxx.2 classifications apply to the initial ingot. The recognized industry specifications allow for some loss of magnesium and some gain in iron from the process of final casting. Final cast alloys should meet the xxx.0 classification and should not be held to the xxx.1 or xxx.2 requirements for alloying constituents. For example, the alloy 356.0 and alloy 356.1 are essentially the same with the exceptions in boldface in the following table.

Alloy	Si (%)	Fe (%)	Cu (%)	Mn (%)	Mg (%)	Zn (%)	Ti (%)
356.0	6.5–7.5	**0.6**	0.25	0.35	**0.20–0.45**	0.35	0.25
356.1	6.5–7.5	0.5	0.25	0.35	0.25–0.45	0.35	0.25

As you can see, the amount of allowable iron is greater in the final casting to allow for pickup of iron in the process, and the amount of magnesium has a wider range limit at the lower end to allow for some loss of magnesium in the casting process.

From here, the system becomes somewhat complex and confusing. An alphanumeric code can precede the alloy designation to identify specific variations within the alloy type. These are variations of particular alloys that still fall within the chemical ranges of the alloy, but they differ due to processes of casting.

For example, there are several variations of the alloy 356.0. There are 356.0, A356.0, B356.0, C356.0, and F356.0 variations. Each of the 356.0 alloy types fall in the chemistry range of the 356 alloy but each has a slight variation in the range or in the levels of impurities that are allowed. The various alloy types defined by 356 are described in the following table. As you can see, several of the alloying constituents' limits and ranges are different.

Alloy	Si (%)	Fe (%)	Cu (%)	Mn (%)	Mg (%)	Zn (%)	Ti (%)
356.0	6.5–7.5	0.60	0.25	0.35	0.20–0.45	0.35	0.25
A356.0	6.5–7.5	0.20	0.20	0.10	0.20–0.45	0.10	0.20
B356.0	6.5–7.5	0.09	0.05	0.05	0.20–0.45	0.05	0.04–0.20
C356.0	6.5–7.5	0.07	0.05	0.05	0.20–0.45	0.05	0.04–0.20
F356.0	6.5–7.5	0.20	0.20	0.20	0.12–0.25	0.10	0.04–0.20

The UNS System for cast aluminum uses an alpha-numeric numbering system similar to the wrought alloys. It attempts to correct some of the confusion within the alloy designations systems. However, over the years, foundries have become familiar with industry jargon, so it is important to clarify with the foundry what specific parameters they intend to work within.

The Unified Numbering System (UNS) System

The UNS System begins with the letter "A," as with the wrought alloy description. The next letter corresponds to the old system, where a 0 is used if there is no preceding letter. Number 1 is used if the preceding letter is A. Number 2 is used if the preceding letter is a B, and so forth up to number 6 if the preceding letter is F.

AA system	UNS system
356.0	A03560
A356.0	A13560
B356.0	A23560
C356.0	A33560
F356.0	A63560

The designations are confusing In the United States because different foundries and different sources adhere to variations of the old systems that have been used over the past century. In the United States, the Society of Automotive Engineers, SAE, has alloy naming criteria that is a variation of the naming used by the Aluminum Association. When the European or Asian specifications are incorporated, there is no end to the confusion. For example, there are more than nine different naming systems for the cast UNS Alloy A13560.

UNS	AA	SAE	ISO	EN AC	UK	France	Germany	Italy	Japan
A13560	A356.0	323	Al-Si7Mg	42 000	LM25	A-S7G	G-AlSi7Mg	3599	AC 4C

These are all essentially the same alloy. It appears that the world of metallurgy has similar communication nuances as the world of language. In Appendix D you can find a cross-reference of some of the common alloys used in art and architecture.

The majority of the alloys used in art in architecture fall into the A93xxx, A95xxx, and A96xxx alloy groups. These have good clarity in appearance combined with superior corrosion resistance.

Cast Alloys

Successful casting with aluminum requires close attention to the percentage of the various alloying elements incorporated in the molten metal as well as trace elements that can contaminate and affect the end casting. Pure aluminum is difficult to cast. It is subject to high shrinkage as it solidifies, so a larger riser is needed as well as high copes to keep the pressure on the solidifying metal.

For many architectural castings, silicon is a major element added to the aluminum. Silicon improves flow but just 0.001% differences in calcium or sodium in the melt can affect the soundness of the casting.

The major factors in determining what alloy to use when selecting the cast form of aluminum is the end quality of the part and the cost. Quality incorporates the functionality and exposure of the casting over a useful lifetime. It also incorporates the surface finish and end appearance. Cost incorporates availability, design of the part, and method of casting (Figure 2.12).

FIGURE 2.12 Upper image is overall cast aluminum *Federal Triangle Flowers*, by Stephen Robin. Lower image is a close-up of the aluminum surface.

TABLE 2.7 Cast alloys of aluminum.

Cast Alloy	Major alloying elements	Characteristics
A03190	Silicon and copper	Good fluidity. Poor corrosion resistant. Sand and permanent mold.
A13190	Silicon and copper	Good fluidity. Poor corrosion resistant. Sand and permanent mold.
A03560	Silicon	Good strength. Good corrosion resistance. Sand and permanent mold.
A13560	Silicon	Good strength. Good corrosion resistance. Sand and permanent mold.
A02080	Silicon and copper	Fair strength. Corrosion resistance is poor. Good sand cast alloy.
A04430	Silicon	Good corrosion resistance. Adequate strength. Sand cast alloy.
A24430	Silicon	Good corrosion resistance. Adequate strength. Permanent mold cast alloy.
A12010	Copper	Highest tensile strength of cast aluminum alloys. Poor corrosion resistance.
A05180	Magnesium	Tough and strong, die cast alloy. Finish imparted from die.
A05200	Magnesium	Tough and strong, sand cast alloy. Will take a good finish.
A05350	Magnesium	Poor casting attributes but finishes well. Sand cast alloy.
A15350	Magnesium	Poor casting attributes but finishes well. Sand cast alloy.
A25350	Magnesium	Poor casting attributes but finishes well. Sand cast alloy.
A07120	Zinc	Good strength. Sand cast and permanent mold cast alloy.
A07130	Zinc	Good strength. Sand cast and permanent mold cast alloy.

Choosing the right alloy is an important first step for the success of aluminum in the various environments. Many variables play into the choice of alloy. Availability, foundry techniques, final finish desired, and intended environmental exposure will determine the appropriate alloy.

Table 2.7 is a list of the cast alloys often used or considered for use in architectural and art when castings play a role. This is by no means a complete list of cast alloys but covers the distinctive alloys used for architectural purposes.

The alloys that are high in silicon with copper and magnesium are good detail casting alloys with good fluidity. These alloys are used mainly in sand casting or permanent mold cast methods. The corrosion resistance of the high silicon alloys is very good. The drawback to the high silicon alloys is the ability to achieve a good surface finish. These high silicon alloys often result in a mottled finish. They can be anodized but the result is often cloudy (see Table 2.8).

84 Chapter 2 Aluminum Alloys

TABLE 2.8 Effects of elements added to aluminum castings.

Element	Range (%)	Effect on the aluminum casting
Pure	99	Pure aluminum when cast will experience high shrinkage.
Copper	5–12	Improves strength and hardness. Lacks fluidity. Hot tearing issue.
Silicon	Up to 12.6	Improves fluidity. Reduces shrinkage. Reduces hot tearing.
Magnesium	Up to 6	Strengthens the casting. Improves corrosion resistance. Dross increases.
Magnesium	>6	Casting can be thermally treated to improve strength.
Iron	Up to 1	Considered a contaminate. Will increase hardness in certain alloys.
Zinc	4–8	Improves strength and hardness. Short solidification range.

Silicon and Silicon Copper Cast Alloys

The most common of the silicon and silicon and copper-bearing aluminum alloys are:

A03190, also known as A319

A13190

A03560, also known as A356

A13560

The silicon aluminum cast alloys are common for most large sand cast and permanent mold cast surfaces because they do not shrink as significantly as other aluminum alloys and because they have good fluidity, which makes it easier to develop detail in the casting.

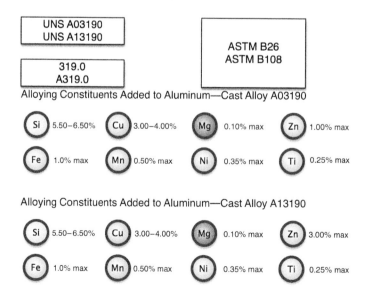

ALUMINUM ALLOY A03190

UNS A03190
UNS A13190

319.0
A319.0

ASTM B26
ASTM B108

Alloying Constituents Added to Aluminum—Cast Alloy A03190

Si 5.50–6.50%   Cu 3.00–4.00%   Mg 0.10% max   Zn 1.00% max
Fe 1.0% max   Mn 0.50% max   Ni 0.35% max   Ti 0.25% max

Alloying Constituents Added to Aluminum—Cast Alloy A13190

Si 5.50–6.50%   Cu 3.00–4.00%   Mg 0.10% max   Zn 3.00% max
Fe 1.0% max   Mn 0.50% max   Ni 0.35% max   Ti 0.25% max

Alloy	Tensile strength		Yield strength		Elongation % (2 in.–50 mm)	Brinell hardness number
A03190-F	27 ksi	185 MPa	18 ksi	125 MPa	2	70
A03190-T6	31 ksi	213 MPa	24 ksi	165 MPa	1.5–3	90

This alloy is a good sand or permanent mold casting alloy from the standpoint of fluidity of the casting. It does not anodize well due to the high silicon content. Corrosion resistance is poor. Note the addition of nickel to this cast alloy. This allows the casting to retain strength at elevated temperatures. Engine blocks and other uses where strength is needed at elevated temperatures utilize this alloy of aluminum.

ALUMINUM ALLOY A03560

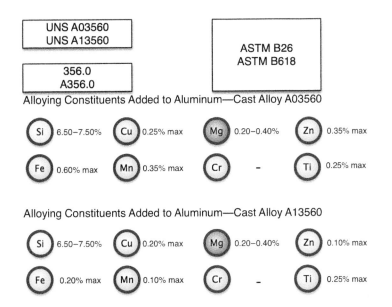

Alloy	Tensile strength		Yield strength		Elongation % (2 in.–50 mm)	Brinell hardness number
A03560-T6	33 ksi	230 MPa	24 ksi	165 MPa	3–7	70
A13560-T6	40 ksi	280 MPa	30 ksi	210 MPa	3–10	75

FIGURE 2.13   Cast aluminum balusters. Alloy A03560, painted white.

This alloy is a good, general duty aluminum casting alloy. It has good fluidity and can be sand cast and permanent mold cast. It does not anodize well due to the high silicon content and the surface finishing character is marginal. These alloys have similar characteristics to the wrought alloy A96061. The corrosion resistance is good in most atmospheric exposures. This alloy can be heat treated to improve strength and hardness (Figure 2.13).

The next set of cast alloys also have high silicon and copper. These alloys have good fluidity, which enables the metal to flow and allows for thinner sections and fine detail. The higher copper content in A02080 reduces the corrosion resistance somewhat. They are generally provided as fabricated temper. The surface finish and the ability to work the surface is not as good as other cast alloys.

Alloy Designation System

ALUMINUM ALLOY A02950

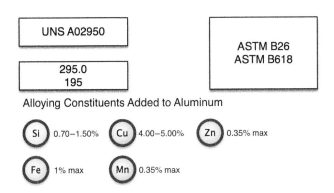

Alloy	Tensile strength		Yield strength		Elongation % (2 in.–50 mm)	Brinell hardness number
A02950-T6	36 ksi	250 MPa	23 ksi	160 MPa	3.8	75

This is a sand cast alloy. Alloy A02950 was a common early cast alloy of aluminum. It was the alloy used as the ornamental aluminum castings on the Empire State Building in New York. Not as common today, this alloy lacks fluidity, so it is more difficult to arrive at good detail. It has good hardness, but the copper makes it less corrosion resistant. Copper was integrated into the early aluminum alloy mixtures because it provided strength and increased hardness. Copper added to aluminum disperses throughout the mixture, giving an even distribution and a homogenous solid solution.

ALUMINUM ALLOY A02080

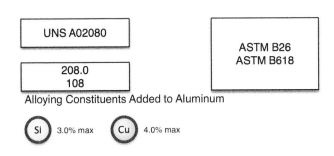

Chapter 2   Aluminum Alloys

Alloy	Tensile strength		Yield strength		Elongation % (2 in.–50 mm)	Brinell hardness number
A02080-F	21 ksi	145 MPa	14 ksi	97 MPa	2.5	55

This is a good sand-casting alloy with good fluidity. It has low strength and fair corrosion resistance. Similar to the A02950, the copper provides some strength and hardness. The silicon improves the fluidity of the metal into the sand cast.

There are modifications to this cast alloy to improve casting into permanent molds. This alloy A12080 increases the silicon to 5.5% and the copper to 4.5%. This improves the welding characteristics as well.

ALUMINUM ALLOY A04430

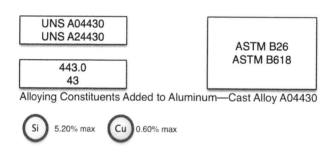

Alloying Constituents Added to Aluminum—Cast Alloy A04430

Alloying Constituents Added to Aluminum—Cast Alloy A24430

Si 5.20% max   Cu 0.15% max

Alloy	Tensile strength		Yield strength		Elongation % (2 in.–50 mm)	Brinell hardness number
A04430-F	19 ksi	130 MPa	8 ksi	55 MPa	8	—
A24430-F	23 ksi	159 MPa	9 ksi	62 MPa	10	—

Alloy A04430 is a sand cast alloy while A24430 is a permanent mold alloy. They have very good fluidity for aluminum castings and excellent resistance to cracking during solidification. They have good corrosion resistance, but they do not anodize well and can be difficult to machine and polish.

The next alloy, A12010 is listed because of its excellent tensile strength. It has a very high copper content and contains other alloys such as titanium to stabilize the grain development.

ALUMINUM ALLOY A12010

UNS A12010

A201

ASTM B26
ASTM B618

Alloying Constituents Added to Aluminum—Cast Alloy A12010

- Si 0.50% max
- Cu 4.00–5.00%
- Mg 0.15–0.35%
- Fe 0.10% max
- Mn 0.20–0.40%
- Ti 0.15–0.35%

Alloy	Tensile strength		Yield strength		Elongation % (2 in.–50 mm)	Brinell hardness number
A12010-T7	70 ksi	480 MPa	61 ksi	420 MPa	4.7	—

Alloy A12010 has the highest tensile strength of the cast aluminum alloys. The high copper content provides excellent strength but only at a sacrifice of corrosion resistance. The T7 temper is normally used with this alloy. T7 corresponds to solution heat treatment processes and is stabilized by artificial aging.

Magnesium Cast Alloys

The next set of cast alloys considered has magnesium as the major alloying constituent. They have decent strength and the surface finish is superior to other cast alloys. Some will accept anodizing well. They are not the easiest to cast, however. They have poor fluidity and will crack if mold design does not accommodate shrinkage during solidification.

ALUMINUM ALLOY A05180

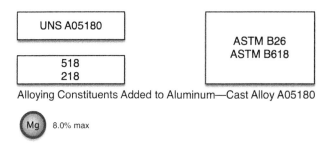

Alloying Constituents Added to Aluminum—Cast Alloy A05180

Alloy	Tensile strength		Yield strength		Elongation % (2 in.–50 mm)	Brinell hardness number
A05180-F	45 ksi	310 MPa	28 ksi	198 Mpa	5	80

Alloy A95180 is a die cast alloy formulated for this process. The high percentage of magnesium increases hardness and toughness of this alloy. The finish of the alloy is induced from the die mold and is usually very good. Post finishing is rarely performed because the die cast mold provides the finish surface quality.

ALUMINUM ALLOY A05200

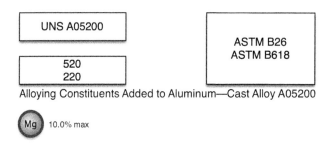

Alloying Constituents Added to Aluminum—Cast Alloy A05200

Alloy	Tensile strength		Yield strength		Elongation % (2 in.–50 mm)	Brinell hardness number
A05200-T4	48 ksi	330 Mpa	26 ksi	180 Mpa	16	80

Alloy A05200 is the sand cast version of the A05180 alloy. The high magnesium content and low silicon makes this alloy harder and stronger, but the casting profile is poor. It has poor fluidity and is susceptible to cracking as it cools. Corrosion resistance is good, and the quality obtained from anodizing is good. It can be machined, and post polishing can improve the surface appearance.

Alloy Designation System 91

ALUMINUM ALLOY A05350

UNS A05350
UNS A15350
UNS A25350

ASTM B26
ASTM B618

535.0
A535.0
B535.0
Almag 35
A218
B218

Alloying Constituents Added to Aluminum—Cast Alloy A05350

Alloying Constituents Added to Aluminum—Cast Alloy A15350

Alloying Constituents Added to Aluminum—Cast Alloy A25350

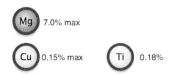

Alloy	Tensile strength		Yield strength		Elongation % (2 in.–50 mm)	Brinell hardness number
A05350-F	40 ksi	275 Mpa	20 ksi	140 Mpa	13	—

Alloy A05350 is a sand cast alloy. It has several refinements to create alloys A15350 and A25350 for specific attributes in the casting process, such as small additions of copper. The base alloy A05300 has boron and beryllium added in very small amounts to thwart oxidation at high temperatures. It is removed in the other forms because of the toxic nature of beryllium.

Alloy A05350 and its sister alloys possess poor casting attributes. Fluidity and susceptibility to cracking during solidification make these difficult to cast. The corrosion resistance is superior, as is the machinability, finishing ability, and quality of anodizing.

Zinc Cast Alloys

These last two cast alloys have zinc as the primary alloying element. The zinc improves the strength, but the casting properties and finishing ability are not as good as other cast alloys.

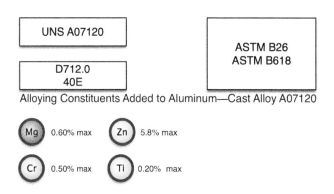

Alloy	Tensile strength		Yield strength		Elongation % (2 in.–50 mm)	Brinell hardness number
A07120-F	35 ksi	240 MPa	25 ksi	170 MPa	5	75

This is a sand or permanent mold cast alloy. The zinc improves strength. Corrosion resistance is average, and the cast properties are poor in comparison to other alloys that contain silicon.

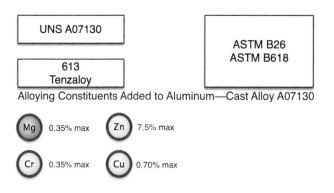

Alloy	Tensile strength		Yield strength		Elongation % (2 in.–50 mm)	Brinell hardness number
A07130-T5	30 ksi	210 MPa	22 ksi	150 MPa	3	75

This alloy can be sand or permanent mold cast. The high amount of zinc will aid in achieving good strength. It can be heat treated to improve strength. It is difficult to weld, and corrosion resistance is average. Anodizing will tend to appear yellow due to the chromium. It can be polished and machined.

These are the alloys of aluminum considered for many art and architecture projects. A few of the alloys listed are rarely if ever used in art or architecture but they possess certain characteristics that might eventually be employed to meet design requirements.

As you explore aluminum for use in a particular application, it is important to check on availability with the mill, the foundry, and the supply distribution house. Some alloys may not be available or might have long lead times. The market will dictate the cost and availability of all forms and alloys, as will the quantity. If the quantity is sufficient and the necessary lead time is in place, most mills will cast the ingot for the precise alloy. The mill or a secondary processor will produce the end form in the required temper and finish.

If the quantity is not sufficient for the mill to make an ingot cast, then you will need to explore availability from secondary houses that inventory the various forms of aluminum. Many of the alloys listed in this chapter are common and often stocked; however, the specific thickness of sheet or plate needed may not be stocked.

CHAPTER 3

Surface Finishing

A man should look for what is, and not for what he thinks should be.

—Albert Einstein

One of the intriguing aspects of working with aluminum is the way the metal interacts with light. Because of aluminum's clear oxide, which forms instantaneously on the surface, and the natural corrosion resistance that this oxide affords, the natural, uncoated appearance can be left exposed in many applications. This natural surface can be polished mechanically and chemically, brushed and etched, or it can undergo a transition by anodizing.

Silver and aluminum share a close relationship in their ability to reflect light. Both reflect the visible spectrum at wavelengths above 350 µm. Pure aluminum approaches 90% reflectance of the light wave while pure silver approaches 98% of the visual spectrum. Aluminum reflects light consistently across the visible spectrum. When the surface of aluminum has a diffuse texture, the reflectivity has a white hue because of this consistency (Figure 3.1).

One early use of aluminum that continues today is as a reflector. The surface provided from the mill in the specular finish will reflect approximately 53% of light. Buffing the surface to a mirrorlike finish pushes this to approximately 72%, whereas chemical brightening processes push reflectivity close to 90%.

A designer working with aluminum is afforded a number of options. Aluminum used in art and architecture, regardless of form, is more often modified with a conversion oxide coating, known as anodized, or coated with a paint layer. Both of these treatments can provide color to aluminum. Paint coatings are varied and include a pigment for color. Aluminum is ideal as a substrate for paint coatings due to the corrosion resistance of the metal. If the paint is scratched and the aluminum is exposed, rust will not present itself. Clear aluminum oxide will form, sealing the scratch.

Anodized conversion treatments impart color by dyeing the porous oxide surface with organic dyes or with metallic compounds, then sealing the pores on the surface. Aluminum can also achieve

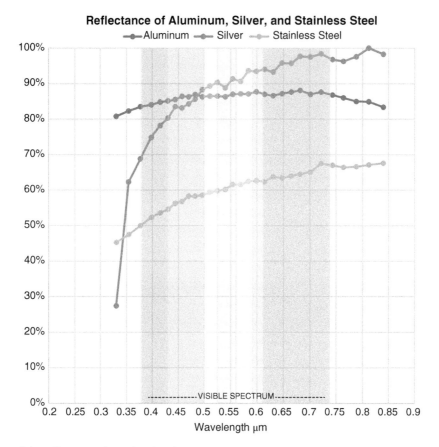

FIGURE 3.1 Light reflectance along the visual spectrum.
Source: Data from NASA.

color from chemical reactions with the alloying constituents in the metal, a method that is used less frequently today.

THE MILL FINISH SURFACE

The plate and sheet mill, extrusion press, or casting facility produces the aluminum regardless of alloy, in what is often referred to as the "as fabricated" state. This is not to be confused with subsequent fabrication processes performed on the aluminum. The "as fabricated" finish is the initial finish provided on the aluminum at the mill, foundry, or extrusion facility. This is the basic raw form. It is also referred to as "the mill finish surface." This is the most economical surface finish and precedes other processes that may involve shaping, welding, and stamping. Pretreatments for paint applications will be performed on the mill surfaces, as well as subsequent finishing and polishing processes.

This mill surface finish, if handled correctly both at the mill and at the subsequent fabrication facility, can be the final finish. Many surfaces are installed in the mill state. More often, however, finishing and polishing processes are needed to display the natural color of the metal in a refined, unblemished way (Figure 3.2).

Mill finish surfaces are created at the production mill that turns the cast block or billet into the form that will be further worked in subsequent finishing and fabrication processes. The form may be plate, sheet, wire, or extrusion. Castings are dependent on the quality of the mold that the aluminum is cast into. Sand casting, die casting, and investment casting all will produce a finish on the surface that may or may not be the final surface finish.

The mill finish on extrusions depends on the quality of the die that the aluminum is extruded through and the tempering process that follows. In the mill finish, in a fabricated state, the extruded surface can have lines and streaks running the length of the aluminum.

For sheet and plate forms, there is typically one prime sidethat has a surface that is more refined than the other, caused by the cold rolling process. As the sheet is passed through rolls to reduce its

FIGURE 3.2 Aluminum alloy A93003 with an Alclad coating. After more than 35 years urban exposure the surface appearance is somewhat duller, but still in superior condition.

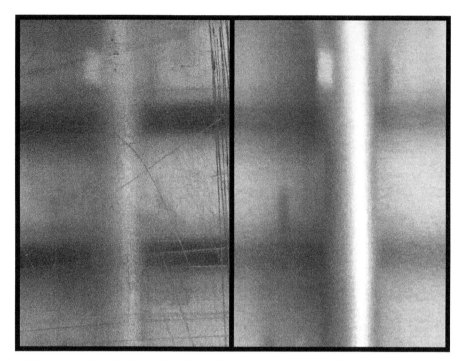

FIGURE 3.3 Nonprime side and prime side of typical aluminum sheet. Not all nonprime sides are scratched to this level. Reflection is a duct in the ceiling.

thickness, the top roll creates a better, smoother surface (see Figure 3.3). Not all surfaces are as scratched as the one shown in Figure 3.3, but the point is that not much care is expended to protect the reverse side of a sheet or plate unless there are very specific instructions.

To obtain a good finish on both surfaces you need to work with a specialty polishing facility and order the metal with a prescribed surface finish on the reverse side. If a sufficient quantity of aluminum is ordered, the mill can achieve a surface of good quality on both sides of the sheet or plate. Unlike the rolling of steels and stainless steels, the back side or underside of sheet and plate aluminum are usually in good condition and free of scratches, mars, or coining streaks from one sheet passing over the edge of another. This is dependent on the coiling and decoiling processing facility because the softness of the aluminum surface requires more delicate handling than is used to produce steel sheet and coil. This leads to a more refined surface on the back side of aluminum sheet or ribbon that makes up a coil.

Aluminum plate with thicknesses greater than 5 mm (0.25 in.) is hot rolled. It can be dull and streaked. Usually there is a cold pass to impart an even, smooth surface. Plates can have Lüder lines, pits, and inclusions. Working with the mill and qualifying the mill finish is critical to achieve a good surface. When ordering small quantities, less than a mill run, the supply warehouses will be the more likely source for the metal. The designer will be constrained to use the aluminum that is in stock or can be obtained readily. Therefore, choices may be limited (Table 3.1).

TABLE 3.1 Mill surfaces.

Mill surface finish	Form	Description
Unspecified	Plate, sheet, casting, extrusion	Mill oil, lines, steaks, inclusions can be apparent
Nonspecular	Plates, sheet, casting, extrusion	Even surface finish, matte appearance
Specular	Sheet and plate	Cold rolled, bright, and reflective surfaces

When mill surface finishes are exposed to the atmosphere, occasional cleaning should be performed. Over time, there will be some decrease in the brightness on the aluminum. Exterior details should allow for the surface to drain and not collect and hold moisture. Where water is allowed to stand, a darkening will occur with some whitish, blurred amorphous deposits along the water line.

Because aluminum is rarely used in a pure state, except as a clad coating, there are intermetallic elements in the alloy matrix. Finishing and polishing can bring some of these to the surface. They are harder than the aluminum, so they are more prone to remain on the surface. These intermetallics—copper, magnesium, silicon, and manganese—can form oxides of their own on the surface, intermixed with the aluminum oxide. In corrosive atmospheric conditions, these can develop into small localized corrosion cells where the intermetallic is the cathode. These can develop into more severe corrosive conditions, but in practice, most environmental exposures never see this. If the surfaces are well drained and an occasional cleaning is performed, there should be minimal problems.

Interior uses of uncoated, natural aluminum will show fingerprints. Fingerprints will eventually etch the surface if they are not removed. Coarse finishes make removal of fingerprints difficult and can require cleaning the entire surface in order to achieve an even finish appearance (Figure 3.4).

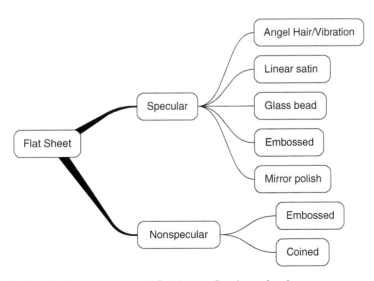

FIGURE 3.4 Mill finish and subsequent natural finishes on flat sheet aluminum.

For flat sheet and plate, the mill provides the aluminum in one of two standard surfaces, specular, and nonspecular. Specular surfaces are clean and smooth, free of scratches or mars. Specular surfaces are passed through clean polished rolls. The top side of the sheet or plate is considered a high-quality specular surface while the bottom surface is "as fabricated," which means just that, as fabricated and passed through the mill rolls (Figure 3.5).

The Aluminum Association lists four mill finishes:

1. AA M10—Unspecified
2. AA M11—Specular as fabricated
3. AA M12—Nonspecular as fabricated
4. AA M1x—Other

FIGURE 3.5 *The Flying Robot,* by Chris Duh. Mill finish aluminum of various forms.

The unspecified surface is not recommended for art or architecture. The quality is usually very poor, and the surface can be scratched and marred to the point that refinishing is a futile exercise. The surface quality should be specified on the order and discussed if necessary. "As fabricated" is a mill term that refers to sheet, plate, or other forms created at the mill, casting facility, or extrusion facility. It does not mean the fabricator of the finish panel or assembly. The "as fabricated" finish would be the finish on most extruded sections and aluminum tubing or drawn pipe. The "as fabricated" finish for plate is typically a cold pass that is smooth but not a finish surface. On bar and rod the "as fabricated" finish can be rough and have dark mar lines.

The sheet or plate can remain as provided from the mill or directed to another facility, where secondary finishing and buffing operations can be performed. This is typically how the metal is furnished to the marketplace. The mill may produce the sheet, but distribution of the sheet, plate, tubing, extrusion, or wire may be by another source. Similar to stainless steel, these aluminum finishes come in different levels and depths. We will use stainless steel finishes for basic comparisons since these finishes and their respective designations are better known in the architectural and art world. The aluminum finish designations are not as broad as the finishes applied to stainless steel. There are various finish levels on aluminum, defined by the particular finish grit, speed of belt or wheel, and directionality. A No. 3 finish on stainless steel is only a close representation of a coarse linear finish applied on aluminum. A No. 8 mirror finish on stainless steel is simply a mirror finish on aluminum. It is recommended to work with samples provided by the finish applier (Figure 3.6).

FIGURE 3.6 Custom desktop finish, custom radial finish and glass bead finish.

DIRECTIONAL FINISHES

The Aluminum Association provides information on several levels of directional and non-directional finishes. They qualify the finishes as:

Directional Finishes

AA M31 Fine satin	Similar to the level of a fine No. 4 in stainless steel
AA M32 Medium satin	Similar to the level of a medium No. 4 in stainless steel
AA M33 Coarse satin	Similar to the level of a No. 3 finish in stainless steel
AA M34 Hand rubbed	Similar to a hairline finish in stainless steel
AA M35 Brushed	Similar to an Angel Hair or Vibration type finish in stainless steel

NON-DIRECTIONAL FINISHES

For the non-directional finishes, the Aluminum Association designates seven levels of sand and steel shot finishes. These are categorized by the mesh size of the sand, blasting gun distance, and air pressure. They go from extra fine matte finish, AA M41, to a coarse matte finish, M44. Further, they expand this to steel shot of various sizes for fine shot blast to coarse shot blast.

Aluminum can be glass bead blasted to produce a non-directional satin finish, but the surface is matte in appearance and will readily fingerprint, making it difficult to clean. This surface should be immediately sealed or anodized. When anodized, if the bead blasting was evenly applied, the end finish is very clear and has a depth beyond that of clear anodizing without the glass bead blast (Figure 3.7).

When the aluminum has been glass bead blasted and clear anodized, the color and depth is striking. It is different than simply anodizing the surface of a specular aluminum. On the Modern Art Museum in Fort Worth, Texas, designed by Tadao Ando, the surface of the aluminum plates was cleaned and degreased, glass bead blasted, and clear anodized.

Another technique using water combined with abrasives will produce a very fine non-directional matte finish. Clean water, combined with fine grit abrasives, 1000–5000 grit, applied at a constant pressure can produce this special surface finish. The surface must be immediately dried to remove any water from the surface—otherwise, darkening and blotchiness will appear. This technique is called Wet Blasting or Aqua Blasting. The finish produced is lighter than that for glass bead and has a light smoothing nature to the finish versus the relative coarseness of glass beads or the severe coarseness produced by other abrasive blast methods.

FIGURE 3.7 Modern Art Museum by Tadao Ando. Custom A95xxx alloy, glass bead, and clear anodized.

Linear satin, angel hair, or vibration finishes will also readily fingerprint, in particular when the finish is newly applied. New finishes can be sealed with clear lacquer or wax, and can be anodized, but the etching process that precedes anodizing will remove some of the finish. Leaving the surface

FIGURE 3.8 Custom finish applied over mill aluminum. Note the white tone of the reflection. Beth Nybeck, artist.

as it is will require keeping it clean until there is time for a natural oxide layer to form. If a surface is left uncoated, it is recommended that it be kept free from handling until moisture from the atmosphere is absorbed and the surface stabilizes (Figure 3.8).

Interior surfaces will fingerprint easily and readily if left unprotected and exposed to handling. The satin finish produces a series of fine linear scratches that capture and hold oils and dirt. When cleaning these surfaces, some of the surfactants used to offset the dirt and oils will remain in the tiny scratches. If not removed these can cause the aluminum to oxidize, especially if the contaminants have an alkaline makeup. Isopropyl alcohol can offset some of this damage and will dry quickly. Deionized water also works well (Figure 3.9).

MIRROR FINISH

Aluminum can be mirror polished. The surface is bright, and the purer the alloy, the better the reflective clarity. The soft A91xxx alloys are the purest. There are high purity A95xxx and A96xxx alloys as well that when polished, provide a silver-like appearance.

Mirror Finish 105

FIGURE 3.9 *Light Filter*, by Rachel Thomas. Angel Hair over mill aluminum.

Due to the softness of aluminum, sometimes buffing is all that is needed, and one can avoid the intermediate steps of polishing. This requires a good quality surface to begin with (Figures 3.10 and 3.11).

Alloy	Aluminum %	Major added element
A91100	99.00	Silicon
A95005	97.15–97.75%	Magnesium
A96063	97.65–98.5%	Silicon and Magnesium

FIGURE 3.10 Mirror finish on 6 mm aluminum, by Ricardo Regazzoni. Doha, Qatar.

Many reflectors used on light fixtures use polished aluminum to increase the level of light reflection. These reflectors use an alloy of high purity, usually an A91xxx series alloy. Higher purity alloys are softer but a clear surface reflection that covers the greatest range of the visible light spectrum is paramount for light reflectors. Aluminum alloys that offer more strength can be coated with one of the pure aluminum alloys using a process called alclad. Alclad, with a high pure alloy, will give a mirrorlike, clear surface reflection (Figure 3.12).

Aluminum sheet and plate that are to be mirror polished require special alloy control from the mill provider. Specular finished sheet and plate from the mill has a superior quality surface provided

FIGURE 3.11 Detail showing how the mirror aluminum forms were created. Aluminum extrusion was used to fix the milled edges together.

from the mill. It is free of inclusions and has a good consistent appearance. You should request that the mill not print its information on the surface by roller. (See Figure 5.5.)

Subsequent polishing operations can be performed on the aluminum surface followed by buffing. Buffing will involve various polishing rouges. Once these are applied there is a need to remove the residual powder from the surface. The softness of the aluminum surface allows for a good flow

FIGURE 3.12 Alclad over A93003 alloy, after 20 years of exposure. Designed by Antoine Predock. Source: Tampa Museum of Science and Industry.

as the buffing operation is underway. This produces an even, clear, reflective mirrorlike surface (Figure 3.13).

Extruded plates and sections can be difficult to buff and polish due to lines in the aluminum surface. These lines are transferred to the aluminum from the die as the aluminum is extruded through the opening. Minor imperfections are transferred to the surface and removing them can be difficult. You can consider leaving them and polishing over them. This will help conceal the lines to a point. If the intension is to produce a fine mirror polish on an aluminum extrusion, the effort should be made first to the extrusion die itself. The edges of the die opening should be as clean and smooth as possible. This will make the subsequent buffing easier.

Once the mirror polish is satisfactory on the aluminum surface it must be maintained. The mirror finish will fingerprint and can tarnish. Periodic cleaning will be necessary to remove light tarnish and fingerprints from the surface. If the fingerprints have remained on the mirror surface

FIGURE 3.13 Mirror polished aluminum plate, by Jan Hendrix.

for a period of time, you can expect a slight surface etching and a polishing paste may be necessary to bring the surface back.

Uncoated textured surfaces, both embossed and brushed and mill surfaces will darken slightly with time as the natural oxide thickens and combines with pollutants from the air. The darkening will be greatest where moisture or other materials are in more frequent contact. Areas where handling of the surface is more prevalent will also darken more rapidly (Figure 3.14).

BRIGHT DIPPING AND ELECTROPOLISHING

Thin aluminum can also be "bright dipped" to produce a very reflective surface. Sometimes referred to as "bright dip anodizing" because bright dipped aluminum is often anodized to harden the surface. It is a much more involved process than simply anodizing due to the brightening chemistry and the surface action required. Bright dipping is an electrochemical process that smooths the surface of aluminum. Similar to electropolishing, the surface is smoothed at the microscopic surface level.

FIGURE 3.14 Mirror polish over A96005 Alloy, by Jan Hendrix.

Bright dipping produces a deep luster and high reflectivity, and works best on hard temper versions of the alloy. The process is a micro-smoothing process. It will not remove scratches or edge burrs. Alloys of aluminum that provide the best results when bright dip anodized are A91xxx, A95xxx, and A96xxx in the harder tempers.

Bright dipping involves clean, smooth aluminum and specialized processing tanks. The tanks are designed to capture all the gases that are released during the bright dipping process. The aluminum is immersed in a tank containing a mixture that includes phosphoric and nitric acid. As the current is introduced into the aluminum, nitrogen dioxide and nitric oxide are released. These must be collected and should never be released into the atmosphere. Proper aluminum bright dipping systems need a complete hooded capturing system to contain the gas being released. There are additives that reduce the amount of gas but a properly functioning system is still necessary to capture the gases.

ELECTROPOLISHING ALUMINUM

Aluminum can be electropolished to brighten the surface. Highly pure forms of aluminum are necessary for adequate results. Similar to brightening, the process removes the high points on the metal surface by concentrating an electrical current at the points that extend out from the metal surface. The aluminum part is made the anode in an electrolyte. Aluminum is amphoteric, so either acid or alkaline electrolytes are possible. A direct current is applied, and metal ions are removed at a controlled rate. Current flows from the anode, which becomes highly polarized. Metal ions diffuse through the electrolyte to the cathode. This removes metal on the anode at a controlled rate. Current and time are the variables in the process.

Electropolishing has a number of beneficial effects on the surface of metals. The finish is nondirectional. The process works on the entire surface at roughly an equal intensity, unlike buffing or polishing, which has a smearing effect in the direction of the tool on the surface and can be subject to localized high and low points. The surface of the metal is not stressed from the operation of buffing and polishing, so shaping from differential stresses will not occur. Shaped parts can be electropolished as well as flat. The process of electropolish floods the surface of the part with small oxygen bubbles, which pulls the hydrogen away. Surfaces devoid of hydrogen will not support the growth of bacteria and this will produce a fine, reflective surface for future processing.

Electropolishing of aluminum is not widely practiced. The majority of electropolishing operations are set up for stainless steel, so finding a facility that has the setup for aluminum and is willing to perform the work may be difficult (see Table 3.2 and Figure 3.15).

ANODIZING

Aluminum quickly forms an oxide on the surface when exposed to the atmosphere. This oxide is very thin and tenacious but enables aluminum to resist corroding substances in the environment. Nearly all metals will react with oxygen and form oxides when exposed to the atmosphere. Metals will also form an oxide when confronted with a strong electrolyte and subjected to an electrical current, in which the subject metal is made the anode or positive polarity. Aluminum, when placed in a powerful electrolyte and exposed to an electrical current in such a way that the aluminum

TABLE 3.2 Basic aluminum finishes.

Finish	Limitations
Mill finish—as fabricated	The finish often provided on extrusions, tube and pipe and castings. It can have mars and scratches. Should not be considered for plate or sheet used on art and architectural purposes.
Mill finish—nonspecular	This finish is available on sheets and plates. It is a glossy surface but not as reflective as the specular finish. Common finish on aluminum destined for additional finishing processes, such as polishing and anodizing.
Mill finish—specular	This finish is reflective and high gloss. This is a good base for anodizing and mirror polishing.
Mirror polish	Available on sheet or plate, both surfaces. Extrusions or casting can be electropolished to brighten but true mirror finish requires polishing and buffing the surface.
Satin-directional polish	Available on castings, extrusions sheet or plate, both surfaces. Some porosity may show on cast surfaces. Complex cast surfaces will require hand finishing which could lead to mottling.
Satin nondirectional polish-Angel hair	Available on castings, extrusions sheet or plate, both surfaces. On complex shapes this will lead to a mottling appearance.
Satin, non-directional polish-Glass bead	Available on castings, extrusions sheet or plate, both surfaces. Caution, very susceptible to variations in blast process. Exercise tight controls.
Embossed	Sheet only.
Custom embossed	Sheet and plate.
Hammered, needle hammered	Thick sheet and plate, castings.
Etching/chemical milling	Sheet, plate, casting, or extrusion.

is made an anode, will develop a thicker oxide with special characteristics. This highly controlled process, commonly referred to as anodizing, can thicken this oxide layer across the entire surface, front, back, holes, and edges. The ability to control and enhance the growth of this incredible oxide is one of the most remarkable properties of aluminum.

Anodizing is an electrochemical treatment performed on aluminum of any shape or form. Anodizing is one of the most common surface treatments used on aluminum in art and architecture. Along with a pleasing appearance, anodized aluminum is impervious to saltwater, dielectric, nearly ceramic in nature, harder than the base metal, and increases emissivity[1] of light. Anodizing

---

[1] Emissivity is the ratio of total radiation given off by a surface to that of a perfect radiator (black box). For aluminum the emissivity is 0.02 and for anodized aluminum it is approximately 0.11, which is similar to silver.

Anodizing 113

FIGURE 3.15 *Rules of Basketball*. A96061 T6 plate, Angel Hair finish over mill surface and machined lettering cut into aluminum plates.
Designed by Gould Evans.

develops a thick, porous oxide. The oxide that develops is polymorphic: the initial oxide is an amorphous surface and the outer, porous oxide is highly crystalline. The makeup of the oxide is similar to alumina, Al_2O_3, and is very stable and hard, similar to the mineral corundum. The pores created as the oxidation develops can be filled with substances that impart rich color into the aluminum or they can be sealed, leaving a matte gray appearance referred to as clear anodize.

Attributes Derived from Anodizing
- Pleasing appearance
- Uniform and consistent surface
- Color
- Improved corrosion resistant
- Dielectric coating
- Surface hardness
- Reduces friction and galling
- Inert surface
- Increased emissivity

There are several electrolytes used to create the anodic coating on aluminum but the most common one used in art and architecture is dilute sulfuric acid. With the sulfuric acid electrolyte, bright finishes are also possible. The surface of the aluminum before immersion in the electrolyte can be brightened by electrolytic or chemical treatment. This is common in light reflectors. On very pure forms of aluminum, 99.99% pure, the brightened surface can achieve a reflectivity of nearly 90%. Untarnished silver is the only material that is more reflective (Figure 3.16).

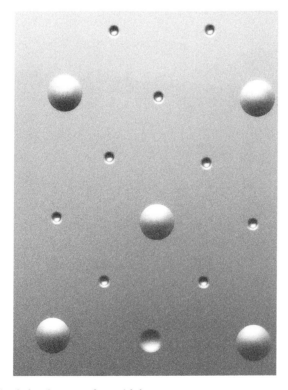

FIGURE 3.16 Clear anodized aluminum surface with bumps.

HISTORY OF ANODIZING

The anodizing process was first developed in 1920s to resist corrosion of airplane struts. Seaplanes constructed of Duralumin aluminum alloys corroded quickly in certain environments, particularly marine environments. The process of anodizing the aluminum alloy helped resist corrosive effects and prolong the service life of the early airplane. The Japanese used oxalic acid as the electrolyte and patented an anodizing process in 1923. After World War II, research on anodizing processes continued in Japan, particularly in the use of color anodizing. Dr. Tahei Asada introduced various metal salts into the anodizing bath and used alternating current rather than direct current to produce his finishes. He applied for a patent in the early 1960s. The Asada process is the basis for the "two-step" anodizing process used extensively today.

The early use of sulfuric acid in the anodizing process was patented in 1927 in Great Britain. This process, which introduces dilute sulfuric acid as the electrolyte, is the most common method used to anodized aluminum. The sulfuric acid electrolyte uses a direct current power source to create the porous aluminum oxide film.

In the early 1930s, color anodizing was advanced in Italy by Dr. V. Carboni. Dr. Carboni received a patent in 1936 for coloring anodized aluminum by introducing metal salts of copper, silver, and nickel into the pores. Furniture manufactured from aluminum tubing was color-anodized in the 1930s using his methods. The method of adding dyes into the anodized surface continued to develop after World War II. Italian designs incorporated colors into common household goods.

From an architectural and art perspective, anodizing did not come into significant use until the 1940s, after the war. The war effort built up a massive capacity for aluminum production and manufacturing and it now needed another customer. This ability to produce aluminum in its various forms presented possibilities to the design world and led to the creation of all types of products developed from the metal. The anodizing process was key to the development of desirable and inexpensive household goods. In the postwar years of the 1940s and early 1950s the bright colors achieved by dye anodizing were used to produce tumblers and pitchers for household use. At the time, two companies in Italy, Bascal and Sunburst, produced these items and sold them worldwide. Heller Hostess Ware was a product line from the late 1940s into the 1950s that manufactured numerous brightly colored tumblers and pitchers. Called the "Colorama" series, these exemplified the postwar sentiment of better times. The products came in various shades of red, blue, gold, and green. These used organic and inorganic dyes to impart the color. The National Museum of American History even has them on display as an example of American life in the late 1950s and early 1960s (Figure 3.17).

Anodic coatings are widely accepted as premier architectural metal treatments. Aluminum is well suited for anodizing and there are numerous firms that specialize in the process. All forms of aluminum can be anodized, within the limitations caused by tank dimensions, depth, width, and length, as well as the dimensions of the part itself. Many plants are set up to process the aluminum parts in mass quantity and there are batch plants with slower unit-by-unit processes that also make parts, but usually at a higher unit cost.

Anodizing involves immersion of the aluminum sheet, plate, extrusion, or form in an acid electrolyte as a direct current is applied. Some of the aluminum dissolves while most of it develops a

FIGURE 3.17 Heller Hostess Ware Cups.

porous cellular oxide. These cells and the pores within the cells are very small and tightly spaced. These microscopic pores are the heart of the anodizing process.

Anodizing is a conversion of the surface, not a deposition or coating in the sense of applying something over the base aluminum. The surface of the aluminum is converted to a thick aluminum oxide that grows on the surface and into the surface. In other words, if the thickness of the oxide coating results in a thickness of 10 μm, approximately 60% or 6 μm grows outward from the surface while 4 μm grows inward into the surface. In conventional anodizing, where sulfuric acid is used, some of the aluminum is consumed as it dissolves in the electrolyte solution. The aluminum that remains on the surface combines with oxygen in the solution to form the oxide layer. The oxide film that develops is both amorphous Al_2O_3 and crystalline Al_2O_3 with a small amount of hydrated oxide on the outer surface. This hydrated oxide is similar to the mineral böhmite ($Al_2O_3 \cdot H_2O$).

The aluminum oxide formed during the anodizing process is a very hard, crystalline substance that resembles corundum in hardness. Corundum is often used as an abrasive to impart brushed finishes on stainless steels and other metals. The crystalline surface that grows on the aluminum is porous. One process, not used in art and architecture, called hard-coat anodizing or hard anodizing, can create a dense oxide surface that can be much harder than steel or stainless steel surfaces.

Substance	Vickers Hv
Aluminum	100–120
Typical anodized aluminum surface	200–300
Steel	200–220
Stainless steel	300–350
Hardcoat anodized surface	350–550

ANODIZING PROCESS

The process of anodizing can be performed on sheets, plates, milled and formed parts, extrusions, and castings. In general, the process involves immersion into various tanks of chemical solutions and rinses. Each step is discrete and critical for success, even the rinsing steps.

Step One: Clean Surface

The first step in the process involves thoroughly cleaning the surface to be anodized. Use of hot baths of alkaline degreasers and detergents are common. Fingerprints, adhesives from protective films, oils, and other organic matter must be removed from the surface. If there are heavy oxides or chemical stains on the surface, they will require mechanical removal first, otherwise they will disturb the ability of the aluminum surface to be anodized. These will not be removed in the cleaning and degreasing process nor in subsequent caustic treatments.

This initial degreasing step can use an alkali or an acid. Strong alkali degreasers can dissolve aluminum. Acid degreasers will not dissolve the aluminum but may need agitation to remove the dirt and grease that has been dislodged from the surface, so that it does not redeposit. Acid pretreatments can, however, be more effective both from a cleaning and a cost standpoint.

After the cleaning step, the surface is rinsed in clean water to remove all excess acids, detergents, or alkalis from the part. The surface is now thoroughly cleaned and ready for etching.

Step Two: Caustic Etch

Once the aluminum has been cleaned, the next step is immersion in sodium hydroxide (caustic soda). This strong alkaline etches the aluminum surface and removes the oxide layer; even damaged anodized surfaces and old anodized surfaces can be removed during this procedure. The use of sodium hydroxide will dissolve small amounts of aluminum in the process, leaving a fine matte appearance. Mechanical pretreatments that physically remove the outer layer of aluminum are an option to the chemical etching process. These can, in certain instances, be very effective in preparing the aluminum surface for anodizing. Typically, though, mechanical operations precede the chemical etching and are used to prepare and even out the appearance of the aluminum surface. Polishing, sanding, and blasting the surface with glass are some mechanical processes used to prepare the aluminum surface before chemical etching and subsequent anodizing.

In some processes, chemical polishing treatments are introduced to brighten the aluminum surface, producing a reflective anodized appearance. This is used on light reflectors and other interior applications, since the anodize coating is very thin.

On occasion a darkish surface can be left on the aluminum after the outer oxide layer has been dissolved. This is called smut and is a residue of the alloying elements that are not dissolved in the caustic bath and become redeposited onto the etched surface. Several of the alloy types contain copper, silicon, and iron as trace elements added for specific mechanical properties. The sodium hydroxide selectively attacks the aluminum and dissolves a small amount of the surface. However, the copper, silicon, and iron in certain alloys are not affected and are left behind on the surface as dark smut deposits. If the smut is not removed the resulting anodized surface is mottled and nonuniform. This metallic residue is only removed by immersing the aluminum part in an acid bath, which must precede any further treatment and anodizing.

The aluminum appearance is clear and metallic after this desmutting step. The surface has little to no oxide and is very clean and even in appearance. Following the desmutting step the aluminum is again rinsed in clean water.

Step Three: Anodizing

Immediately after the rinse, the clean and etched aluminum part is immersed in the electrolyte of dilute sulfuric acid. A direct current is applied, with the aluminum part being the anode and a metal cathode set along the perimeter of the tank. The electrolyte in the tank is kept agitated and the oxide begins to grow immediately. The sulfuric acid electrolyte is maintained at a temperature range of 21–24°C.

The porous oxide begins initially to grow out from the surface to a point where electrical resistance slows down the growth and the oxide then begins to grow into the pores. Because this is an electrical process, all edges develop the porous oxide. As the oxide grows it begins to resist the current, so the current becomes concentrated at areas where the oxide has not fully developed. It is not a perfect thickness across a large surface area. Geometry and tank electrode layout will affect the surface more in one area than another. There are always slight variations in thickness when measured at one end to the center of large anodized parts. The appearance however, should be consistent. The surface of the aluminum has an unbroken, clear anodized porous oxide over all areas that the electrolyte and the electric current were able to access.

If there are covered areas that are protected from the electrolyte, such as under clamps or overlaps, the oxide layer will not grow and there will be a visible difference.

Step Four: Coloring (Optional)

Once the part has developed thick clear porous oxide from immersion in the sulfuric acid tank, it can be left as clear anodized aluminum and move to the next step involving sealing, or it can receive organic dyes or inorganic dyes. Organic dyes involve a saturated solution of chemical dye and water. The newly clear anodized part is immersed in a warm, saturated dye solution and remains submerged for several minutes while the dye is absorbed into the pores. The dye solution is agitated, and the dye enters the pores by capillary action to produce the color. The temperature should be maintained at 65°C. Temperature, dye concentration, and time of immersion are all critical factors in producing an intense color. The dye molecules, in this instance, enter the pores just beyond the surface (Figure 3.18).

FIGURE 3.18 Clear anodized, organic dye anodized, and inorganic dye anodized.

If the coloring agent is an inorganic dye, an alternating current is applied to the anodized part. This drives, or pulls the mineral, to the base of the pores, rather than the mouth of the pores, as is the case with organic dyes. The inorganic dye has a charge that as the current alternates, pulls it toward the base and walls of the pores. Inorganic dyes are more color-fast because they are deeper set into the pores and because they are minerals that are not subject to decomposition and fading from ultraviolet radiation. The color options are limited compared to the organic dyes.

Final Step: Sealing of the Surface

This final step is critical where the anodized aluminum surface is sealed. A seal effectively closes the pores and locks in the color. If no color is applied as with clear anodize coatings, the pores are still sealed. The seal keeps oils and other substances out of the pores. Sealing of the pores involves hydrating the outer region of the pores with a nickel salt. Using nickel will improve the properties of the coating without affecting the color. The seal can be achieved by immersion in hot de-ionized water just short of boiling to hydrate the pores and swell shut the entry, or the pores can be sealed in a hot solution containing nickel acetate. The anodized part is submerged in a hot solution maintained at 85°C and the nickel becomes nickel hydroxide in the solution. This is absorbed at the mouth of the pores and seals the surface. The outer edges of the pores expand and form a hydrated aluminum oxide. Hydrated aluminum oxide is a larger molecule than the aluminum oxide alone. This effectively closes the opening of the pores and improves the qualities of the surface.

Another room temperature sealing process has been used as well. It involves nickel fluoride salts. Again, the pores are sealed but the fluoride does not hydrate the outer surface and does not provide the lasting protection of the sealed and expanded pores. Anodized aluminum sealed with fluoride salts has been shown to fade more readily than when sealed with the nickel salts (Figure 3.19).

THE ANODIZED SURFACE LAYER

The process of anodizing creates a dual layer system of protection for the aluminum. The initial layer that forms is a barrier layer oxide film. This film does not grow. This first barrier layer is thinner and composed of amorphous aluminum oxide. Above this grows the porous layer with a crystalline structure. The thickness is limited by an electrodynamic system as the pores grow. Initially, the oxide layer acts as an electric insulator. As the voltage runs through the layer, small pits develop, which allows electrolyte to move closer in and develops deeper pits and eventually pores. The oxide now grows and expands to form the deep porous structure. Some of the aluminum is consumed in the electrolyte but a large outer section converts to this very adherent hydrated aluminum oxide layer (Figure 3.20).

In the acid electrolyte, the aluminum part is made the anode and a cathode is placed in the tank to complete the circuit. Often the cathode is also an aluminum of A96061 or A96063 alloy. The anodizer maintains certain controls, like area of anode to cathode, to ensure an even growth of the oxide. Temperature, bath concentrations, and time of immersion are other important variables in the process of anodizing.

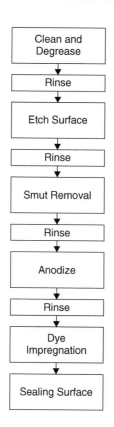

FIGURE 3.19 Anodized process steps.

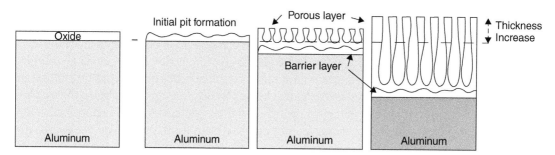

FIGURE 3.20 Aluminum oxide growth from the anodizing process.

The surface of the aluminum is converted to a thick aluminum oxide that grows on the surface and into the surface.

As current is applied, oxygen ions are released from the electrolyte solution and combine with aluminum on the surface. The oxide surface forms rapidly at first and grows inward on the surface and swells outward from the surface. Oxide will continue to develop as long as the resistance to electric flow is not impeded. The oxide itself is a semi-conductor and slows the movement of the current through to the aluminum. The microscopic form of the oxide is a network of tightly formed hexagonal pores. These pores are crystalline in nature at the surface and act as a semiconductor, becoming more and more resistant to the electrical current as the pores grow.

This film has a dual structure: crystalline and amorphous. The crystalline portion has a very regular, hexagonal structure with tiny columnar pores when viewed under a high-power microscope. These pores are near normal to the surface.

At the base of this crystalline honeycomb structure is the amorphous oxide of aluminum. This is a formidable barrier oxide that protects the base metal. The oxide coating is transparent (Figure 3.21).

The refractive index of the oxide coating itself is approximately 1.60–1.65. Compared to water, with a refractive index of 1.33 or polycarbonate with a refractive index of 1.60, aluminum oxide

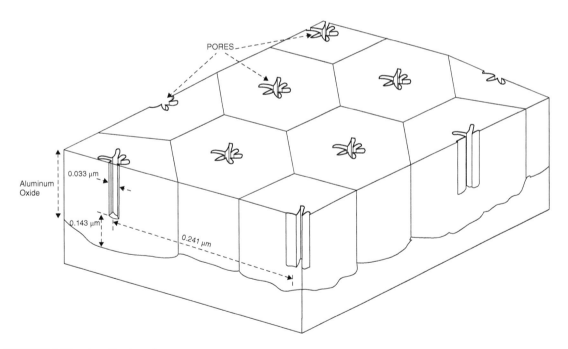

FIGURE 3.21 Anodized surface.

FIGURE 3.22   Microcracks in anodized film.

is very transparent. The oxide created from the anodizing process is harder than the base metal. This hardness makes it more prone to developing small micro-cracks during forming operations performed on anodized aluminum sheet or plate.

For Class I anodized coatings, the anodizing operations should occur postfabrication. On clear anodize sheet these micro-cracks may not be a visual issue. The base material is aluminum and the color of the cracks appears white in contrast to the gray. On anodized sheet with color, the microcracks will be very apparent (Figure 3.22).

All aluminum alloys are suitable for the anodized process. The clearest and brightest anodic films are developed on the purest of the aluminum alloys. The coatings become cloudy and dull as the alloying elements increase. Grain size is another factor influencing the quality of anodic films. The fine grain structure gives the best results. Grain structure is dependent on the tempering process used and the alloy type. Additionally, continuous cast aluminum lacks refined grain size to achieve good color when anodized.

Alloys with Superior Anodizing Quality
A91090
A91080
A91050
A95005
A95083

A95154
A96063

These are the high purity alloys of aluminum. They anodize extremely well because of their higher purity. The oxide film that develops on these alloys is very clear and bright. The trade-off on the A91xxx alloys is that they have very low yield strength.

Alloys with Good Anodizing Quality

A93003
A93105
A95050
A95052
A96061
A96151

Many of these alloys are common in art and architecture. The clarity of the oxide that forms is very good. The surface is not as bright as the high purity alloys due to the alloying elements.

There can be instances with some of the aluminum alloys where the alloying elements, such as copper or silicon separate out near the grain boundaries of the aluminum lattice. These can cloud the oxide and lead to defects in the anodized coating.

Tempering processes, such as cold rolling of sheet or plate, which strain hardens the surface will elongate the grain size and make it more uniform. Further heat treatments to the aluminum can improve the grain size and shape and enhance the appearance of the anodized surface.

Extrusions are produced in long lengths and this makes it difficult to control the tempering and age hardening. Grain size can vary along the same extruded piece and lead to issues with color uniformity. The color issue with extrusions is usually subtle compared to the color differences that occur between flat sheets that have been batch anodized.

Typical anodized surface thicknesses are from 10 μm for interior use to 25 μm for exterior applications. Some hardcoat processes will take the surface to 100 μm in thickness. These are not common in architecture.

For all anodized surfaces, the oxide is clear, but the etching performed before anodizing tends to reduce the surface reflectivity. Etching is necessary to create a uniform surface shade and appearance. Etching has a slight leveling action that can remove minor surface flaws as it removes the oxide on the surface.

Chromic acid anodizing	5 μm in thickness
Sulfuric acid anodizing	5–18 μm in thickness for clear
Sulfuric acid anodizing	10–25 μm in thickness for dye impregnation
Hardcoat process	25–100 μm in thickness

The sulfuric acid electrolyte is the most common process in use. Trade names, such as Alumilite, Anoxal, and Oxydal, all represent similar sulfuric acid processes.

The industry has classified two thickness ranges for architectural anodized surfaces. These are defined as:

Architectural Class I
    No less than 18 μm (0.7 mil)
Architectural Class II
    Minimum 10 μm and maximum 18 μm (0.4–0.7 mil)

The Class I coating is of sufficient thickness to receive coloring processes that are considered more durable and lasting. This would include the mineral dyes or inorganic dyes and electrolytic coloring via the two-step processes. The thickness of these coatings makes them suitable for exterior applications where durability and corrosion resistance are required. It is common to achieve 25–30 μm thicknesses on anodized surfaces used in architecture, in particular for exterior applications. Postforming operations will visibly craze these thicker coatings. Their thickness makes for a ceramic-like surface that will not allow shaping after the process of anodizing.

The Aluminum Association uses Alcoa's designations for the Class I coatings. These are divided further by the anodizing process as follows:

A41	Clear anodized
A42	Integral color anodized
A43	Impregnated dye-color anodized
A44	Electrolytic deposition anodizing

The Class II anodized coatings are thinner. They are suitable for interior applications and some exterior applications where harsh environmental exposures are not a major concern. It is recommended that Class II coatings be maintained by an occasional washing to remove substances from the surface. There have been significant developments to improve the quality of Class II coatings. Coil anodized, composite aluminum panels use anodizing thicknesses that fall within Class II. Many of the coil anodized aluminum sheets and composite anodized aluminum panels are at the lowest thickness of the range, 10 μm. This allows them to be formed after anodizing with less crazing at the bend line but at a sacrifice of the corrosion protection afforded by the thickened oxides (Figure 3.23).

The Aluminum Association's designations for the Class II coatings are divided further by the anodizing process as:

A31	Clear anodized
A32	Integral color anodized
A33	Impregnated dye-color anodized
A34	Electrolytic deposition anodizing

FIGURE 3.23 Class II, coil anodized, used on Windspear Opera House. Designed by Foster Partners.

Studies[2] have shown a direct correlation between the useful life of anodized surfaces and the thickness of the anodic film. The anodic film can be expected to perform well over 10 years in various aggressive environments.

---

[2]V. E. Carter, *Journal of the Institute of Metals* (1972), 100: 208.

Minimum film thickness	Environment
35 μm	Severe industrial environments
16 μm	Moderate industrial environments
12 μm	Marine environments

VARIATIONS IN THE ANODIZED COATING

When anodic coatings are produced, variations in thickness of the finish coating are common and unavoidable. Even across a single sheet or extruded tubing. Thickness variations across a single part can be from 2 μm to as much as 7 μm for the Class I coatings. These variations have to do with a number of variables in the process, such as the electrical current flowing from the cathode to the electrode in the acid electrolyte solution. The solution is kept agitated to maintain a constant and even temperature throughout the solution. The aluminum to be anodized is submerged into the acid solution in such a way to attempt an equal distance between the aluminum surface and the cathodes placed around the tank perimeter. Any aluminum surfaces that are closer to the cathodes will develop a thicker anodic film. Surfaces that are hidden or behind other surfaces will have a thinner coating.

On long thin parts, the center will receive a thicker anodic film than the ends. Sheets and plates may have a thinner anodic film along the edges. With experience and expert instruction it's possible to place the cathodes in a manner that will even out the coating to a degree, but it is impossible to create the same film thickness across the surface of an aluminum part.

Because of this, most specifications qualify a minimum coating thickness. Anodizing facilities will often use a standard deviation calculation to establish a target thickness. These can be specific for a particular project and the particular anodizing plant. It is important to run tests and measure the results in order to arrive at the necessary target thickness. The appropriate standard deviation from the design thickness must be determined and the end result must keep the minimum at or above the specified requirement.

ANODIZING QUALITY

The quality of the anodized aluminum is a function of several variables: the alloy, cold rolling process at the mill, heat treatment, the structure and shape of the grain, and, in the case of extrusions, the roughness of the die.

Variables	Anodize quality issue
Alloy	Etching process leaves smut
Cold rolling	Too little lubrication
Hot rolling	Aluminum oxidation combines with lubricant
Grain	Crystalline structure orientation and size influence appearance

Form a visual context, different alloys will not anodize the same. Tonal differences in color will be apparent even when the same alloy is anodized if the alloy batch is different, that is, it was created at a different heat. The tolerances at the mill allow for range in the alloy consistency and there is simply no way to control this. For example, two mill certificates for A95005 H34 aluminum of the same thickness but cast at different times, several weeks apart, read as follows:

Heat	Si%	Fe%	Cu%	Mn%	Mg%	Cr%	Zn%	Al%
A	0.14	0.50	0.01	0.07	0.85	0.02	0.02	98.38
B	0.12	0.40	0.00	0.01	0.85	0.00	0.01	98.59

Both heat A and B met the standard tolerances for alloy A95005. However, the slight differences of the amounts of trace elements were enough to make one anodized surface a bit darker than the other, even when they were immersed in the same tank together at the same time. Visually, the unanodized sheet was indistinguishable. But once the clear anodized surface was created, there was a visible difference in certain lighting conditions. When a color, in this case a champagne anodized hue, resulted from the mineral dye, the difference in tone remained. See Figure 8.11.

Different tempers will also show differences in color. An H14 temper can have a different appearance than an H34 temper on the same alloy. Any time one surface undergoes a different thermal treatment or a different cold work process, you should expect differences in appearance whether clear anodized or color anodized. The differences may be minor, they may show up only when viewed at a particular angle or under a particular lighting situation.

DEFECTS THAT CAN APPEAR AFTER ANODIZING

The anodizing process is not a covering or a cleaning process. Quality anodizing companies will examine the surface of the aluminum part and treat it to a series of degreasing and cleaning processes, but there are limits to what contaminants can be remove. The etching step dissolves a layer of aluminum from the surface and the appearance is very clean but if there are other substances on the aluminum, they will not be dissolved. The following are a few of the more common substances that can have an effect on the end beauty of the anodized surface.

EFFECTS OF LUBRICANTS

Cold rolling operations involve lubricants to move the sheet or plate of aluminum through the reducing and tempering procedure. When there is too little oil, the aluminum drags on the rolls and this can cause galling. Aluminum is susceptible to galling due to the makeup of its crystal structure. Galling occurs when metal surfaces slide over one another. Friction between the metal surfaces and adhesion on the microscopic level can roughen the surface. Miniscule scratches or burrs can occur on the surface of aluminum sheets and plates. These irregularities will show up more readily after the surface is anodized.

In hot rolling, as the metal is reduced in thickness, lubrication is needed as well. The lubricant can combine with the aluminum oxide. When the surface is anodized, these areas can appear as streaks or blotches in the finish.

STRUCTURAL STREAKS

Grain size control and orientation of the grain is important for a clear, uniform quality anodized surface on aluminum. Clarity, reflectance, and brightness of the surface are affected by the consistency of the grains. Cold rolling establishes an orientation in the grains along the length of the sheet. The grains stretch and become elongated. Casting aluminum develops different sizes of grains during the cooling process. However, some of these grains can vary across a part and, when anodized, they appear as contrasting dark or light blotches or lines running over the surface.

Tempering sheet and cast parts will assist in making the grain size more uniform. Postthermal treatments are critical in achieving clear, uniform surfaces. For extrusions, heat treatment processes are important for surface clarity and consistency.

Extrusions have the additional issue of translating flaws in the dies that the aluminum is extruded through. These will show as lines running the length of the extrusion. The etching process can be prolonged to reduce the lines, but this compounds the maintenance of the etching tanks. A condition known as "structural streaking" can occur with extrusions and will manifest on the surface after anodizing. This condition shows as contrasting lines or streaks running the length of the extrusion and may occur where the extrusion came together and rejoined trapping oxides. These streaks show only after the part is anodized.

MANAGING EXPECTATIONS OF THE ANODIZED SURFACE

There is an assumption when specifying Anodized Quality, or AQ, that the material will appear the same after anodizing, regardless of the batch or lot from which it was obtained. This is not the case. This designation offers no implied warranty that the metal will match in color. It is not possible to produce aluminum alloys at the mill under precisely identical circumstances with identical metallurgical characteristics on different run cycles. The tolerances of alloying constituents exist because it is neither easy nor cost effective, let alone possible, to perfectly replicate the conditions.

The metallurgical structure of the metal also plays a part in color tone. This is true with all-natural metal surfaces. Grain size, heat lines, and flow lines are introduced as the cast ingot of metal is converted to a sheet or plate. These are metallurgic occurrences determined as the metal is converted from one form to another. Grain size is a function of cold working and thermal treatments as well as alloying. Grain size can influence the color tone visible in sheets and plates that are anodized. Adjusting the grain size to a precise match is akin to growing oak trees with identical grain. Flow lines and heat lines are intergranular microstructure conditions that develop as the metal alloy cools and then is rolled thin. These can translate into localized differences in the anodized surface.

Anodized aluminum will produce strikingly beautiful surfaces, but do not expect them to be without variation. Anodizing is a conversion of the surface into a thickened oxide. Colors are set into this oxide or perceived when light passes through and reflects back. The most beautiful metal surfaces depend upon this reflectivity and the variations that the oxide imparts to the reflective tone.

Patchiness should be controlled but occasional tonal differences in a series of anodized panels can add to the beauty of the surface. You should also expect angular differences, like the facets of a diamond, as light reflects off and through the oxide. From one angle the surface or part of the surface may display a certain light or dark tone, while from another point of view the tone changes. This is the character of the natural surface and the way it interplays with natural light (Figure 3.24).

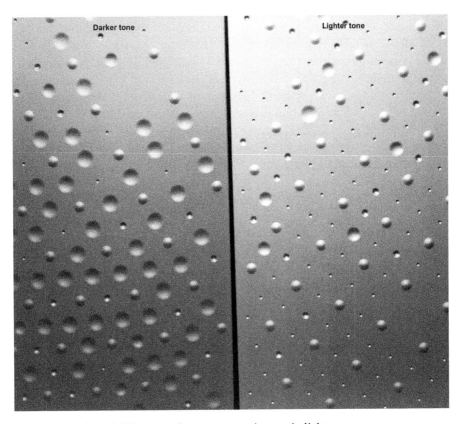

FIGURE 3.24 Example of tonal differences that can appear in certain lights.

COLORING BY MEANS OF ANODIZING

Clear Anodizing

Clear anodizing is the name given to the porous oxide that forms when the aluminum sheet, plate, or extruded part are immersed in the electrolyte of sulfuric acid and a direct current is applied with the aluminum element connected to the positive anode. The longer the aluminum element is in the bath with the applied current, the thicker the porous coating develops, until a point where the resistance from this inert layer negates the current flow.

The clear oxide layer must be sealed in subsequent processes involving a bath in hot water containing a nickel salt. Once this occurs, the surface should display an even, matte, gray-white tone. The matte appearance is due to the etching that preceded the anodizing step in the sulfuric acid.

Clear anodizing is a very common process and produces an even, hard surface. It can be performed on any shape or form of aluminum. Clear anodizing aluminum is an excellent way of producing a surface that resists fingerprinting, staining, and corrosive conditions (Figure 3.25).

Matching in color is subject to variations of alloys and processing at the anodized facility. The thicker anodized coatings will show more variations in the appearance. Because the oxide coating is clear, there are angular differences that are visible when viewing a surface composed of a number of anodized sheet or plate elements. Different batches and different heats of the same alloy will look different in certain lights due to very slight differences in makeup of the alloying constituents. There is no way to control this. The metal should be ordered as "Anodized Quality" and, to the greatest extent possible, should come from the same heat. The pure alloys, A91xxx, produce the clearest and brightest clear anodized coatings. Alloys of the A95xxx and A96xxx also are high in purity and are often anodized. The alloys containing silicon tend to be grayer while those with copper result in dull coatings.

The alloys with lower iron content tend to produce the best clear anodized films. For example, alloy A95005 will give good clarity in the anodic film where alloy A95052, which is higher in iron, will have a slight yellowish tint.

In castings, the alloys with high silicon, alloy A03560 and A03190 will not anodize as well as other alloys with lower silicon levels. Alloys with higher amounts of magnesium, chrome, and zinc—such as A05180, A05350, A07120, A07130—will anodize well but there is a trade-off in the fluidity of the molten metal entering the cast mold.

Hardcoat Anodizing

The process called hardcoat anodizing or hard anodizing produces a very thick anodic coating on the surface of aluminum. It is essentially the same sulfuric acid electrolyte used in most anodizing processes today, but a different technique is implemented to grow a very dense, hard, and less porous surface. The approach is to reduce the amount of aluminum going into solution. A lower sulfuric acid percentage is used at a lower temperature. The electrolyte is agitated as a higher current density is maintained. The resulting surface is rougher but thicknesses from 25 μm to as much as 100 μm

Coloring by Means of Anodizing 131

FIGURE 3.25 Champagne color tone. Designed by José Rafael Moneo Vallés.
Source: Columbia University.

FIGURE 3.26 Bronze anodized extrusions at different angles. Designed by WRNS Architects.

can be developed. It increases the hardness of the metal significantly, however, it is more expensive to produce and takes a significantly longer time and a great amount of energy. The color is gray and the surface is not sealed. It is not suitable for coloring because, as the oxide surface thickens to these levels, it swells in volume. The pores close from this swelling effect and dyeing is therefore not possible.

Color Anodizing

There are several anodizing processes that have been developed to impart color into the aluminum anodized product. Very early in the development of the metal for industrial applications, artistic methods and uses were exploited. These arose from the initial development of the anodizing process on aluminum. It was found that when first anodized, the porous surface is very receptive to dyes. Initially this effect was more of a novelty that found uses in art and architecture; however, it was not long before housewares and utensils were produced with colors (Figure 3.26).

The coloring process was developed in Italy and is credited to Dr. Caboni who patented a process in 1936. He used various metal salts of nickel, copper, and silver introduced into the bath containing the freshly anodized aluminum. The metal salts would deposit into the pores created by the anodizing and impart colors into the aluminum. Caboni would take the freshly anodized aluminum and place it in another bath containing the metal salts. Switching the anode to make the aluminum cathodic, an alternating current was induced. The metal ions would enter the pores and become trapped.

There are five methods used to induce color into aluminum. The most effective methods involve the porosity of the aluminum oxide developed during the anodizing process. Some, however, involve the aluminum alloying constituents, but these are not common in art and architectural uses.

Anodizing process	Vehicle
Dye coloring	Organic and inorganic salts
Alloy coloring	Elements in alloy
Integral coloring	Organic acids
Two-step electrolytic coloring	Metal ions
Three-stage electrolytic coloring	Metal electrodeposition

In each of these methods the metal alloy has some effect on the color obtained but it is the pores that develop on the outer layer of the surface that provide a vehicle for accepting color.

Dye Coloring

Dye color anodizing is a process where the newly anodized surface is immersed in a vat of organic or inorganic dye material. The dye is diffused into the pores and is deposited near the mouth of the pores. After the initial anodizing and before the seal step, the pores of the aluminum are open and deep. When the anodized aluminum sheet is dipped into the dye, the colored pigment seeps into the open pores. The dyed surfaces are immersed in a vat of boiling water where the outer layer of aluminum becomes hydrated and swells. This pinches the opening of the pores and locks the dye into the aluminum.

Organic Dyes

On organic dyes, the colors can be striking and blended to produce intense colors. All varieties of color can be introduced into the pores of the newly anodized sheet. Organic dyes will fade and are not advised for exterior applications. The ultraviolet radiation will eventually break the covalent bonds of the organic molecules of pigments within the dye. There has been some improvement to the early dye formulas that afford longer exposure to the exterior but organic dyes are subject to the same effects that light has on any dye. The colorfast nature is subject to ultraviolet degradation. Remember, on dye-anodized surfaces, the dye is near the mouth of each pore. Protective clear

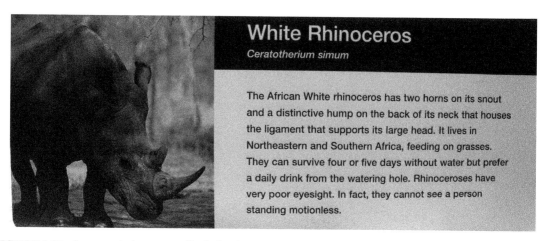

FIGURE 3.27 Image printing on anodized aluminum. These are produced prior to sealing the anodized pores.

coatings can be applied and will provide some ultraviolet protection but at a loss of the metallic tone exhibited by surfaces utilizing the dye alone.

Dyes and pigments by their very nature must absorb light in the visible region of the light spectrum in order to impart color. The dye molecules must be photo-stable under the conditions of practical use as a surfacing material. For organic dyes, the degradation over time and ultraviolet exposure will cause them to fade. Fade is the term given to describe the photochemical degradation of the dye molecules. The eventual fading of the organic dyes and the fact that both organic and inorganic dyes are close to the surface are limitations to this coloring technique (Figure 3.27).

The color range on the organic dyes is striking. Pastels, shades of blues, greens, reds, yellows, and purples as well as deep blacks are possible. The process is not difficult but there are a number of variables to be kept under control for repeated success (Figures 3.28 and 3.29).

Variable	Affect
Alloy	The purer, the better the clarity.
Anodizing thickness	20 μm allows for greater dye absorption.
Time in dye bath	The longer the time, the more dye is absorbed, to a point of saturation.
Concentration	Certain colors require different concentrations.
pH	Maintain pH of 6–7. Some dyes need a lower pH.
Initial finish	Chemical brightening, glass bead, other effects.
Buff after seal	Color buff the surface afterwards to remove traces of seal.

Inorganic Dyes: Pigment Precipitation

With inorganic dyes, the colors are less saturated, but the result is more colorfast. Inorganic dyes are less susceptible to fading. The inorganic dyes are metal salts. These metal salts enter the pores

FIGURE 3.28 *Mixed Media Artist, Artist Proofs*, by Katrina Revenaugh. Ink dye painting on aluminum.

and impart color to the surface of the aluminum. This is also referred to as pigment precipitation. The metallic salts produce a metallic sheen as interference effects are combined with light reflecting from the surface. Gold, blues, reds, and yellows are achievable by the diffusion of metal salts into the pores. This coloring process occurs immediately after the anodizing process and before sealing, similar to organic dye coloring. The aluminum part is dipped in a concentrated solution of metal

FIGURE 3.29 Examples of various colors obtained from dye anodizing. Machined plate, extrusions, solids, and cups and bowl forms.

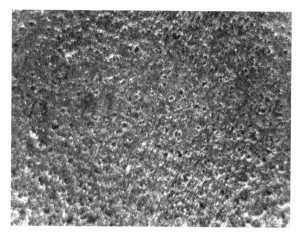

FIGURE 3.30 Microscopic image of dye anodized aluminum.

salts where the color is derived from atomic transitions of the mineral. The metal salts precipitate into the pores and produce a color. The surface is then sealed (Figure 3.30).

The process is not as common today due to the proliferation of the electrolytic process, which also can impart color. The colors produced by the pigment precipitation are:

Yellow—Lead nitrate or lead acetate with potassium dichromate
Red—Lead nitrate or lead acetate with potassium permanganate
Green—Copper sulfate with ammonium sulfide

Coloring by Means of Anodizing 137

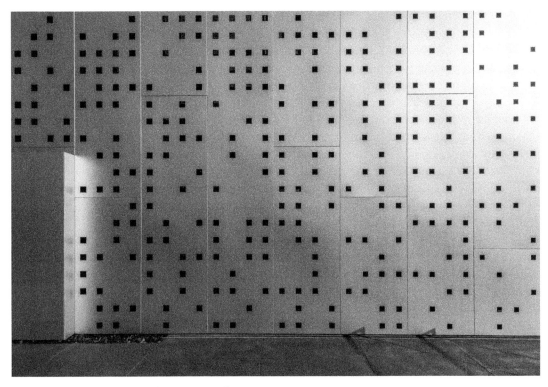

FIGURE 3.31 Gold anodized aluminum used on Haas Museum.
Designed by Stocker Hoesterey Montenegro Architects. Source: Image by Wade Griffin Photography.

Blue—Ferric sulfate with potassium ferrocyanide
Black—Cobalt acetate with ammonium sulfide
Gold—Ferric ammonium oxalate

Ferric ammonium oxalate, for example, is an inorganic salt used to impart a beautiful metallic gold color onto the anodized aluminum surface (Figure 3.31).[3]

Both organic and inorganic dye techniques rely on the pigmented fluids to be absorbed into the pores. In the process, the longer the aluminum is in the dye tank, the deeper the color as it fills the pores. There are diminishing returns as longer time periods will have less effect, but it is important to dye to saturation. For intermediate levels, it is very difficult to impossible to achieve a color match.

---

[3] According to *Plating and Surface Finishing*, 2/79, "the most commonly used inorganic pigment is produced by immersing the freshly anodized part in a 20 g l^{-1} ferric ammonium oxalate solution at pH 5.0 and 50°C; this provides an attractive, metallic gold appearance."

The color intensity for either of these methods depends on the porosity of the anodized film and the clarity of the oxide. The intensity of the color is enhanced by the reflective nature of the aluminum surface. Note that the reflective gloss of the aluminum is dependent on the thickness of the anodizing. As the thickness increases, light reflectance, both total and specular, decreases. A high purity aluminum, however, will retain a higher reflectance even as the thickness increases.

There is a resurgence in the use of color today, but the technique used for exterior applications has moved away from inorganic dye. Today, for better resistance to fading and greater durability, the electrolytic coloring process is preferred. Organic dye impregnation is still extensively used to produce artistic imagery on aluminum and to create very colorful personal utility items, such as aluminum water bottles and other small devices that can be enhanced by the intensive color. Selective printing techniques can produce photographic images on the metal surface or resists applied in the dying process that allow for multicolored affects. Very intricate imagery can be obtained, and can be compared with the clarity of a photograph (Figure 3.32).

You could say we are having a resurgence of 1950s styles, with dye-anodized aluminum becoming a common feature in everyday items. There are several companies that produce a multitude of dyes for use in coloring aluminum surfaces. These dyes are special mixtures, some more durable and colorfast than others. They will fade over time when exposed to light and matching the colors of previous colored pieces can be difficult. One reason for this difficulty in color match lies in the process used originally by the dye anodizing company. The dye used may be known but often the company will alter it slightly, making it a requirement to return to them for matching. Sometimes the dyes would have small additives of other dyes known only by the company doing the color. This would insure they maintained the unique aspect of the coloring. The number and variety of dyes that can be used to color anodized aluminum is extensive (Figure 3.33).

Alloy Coloring

Alloy coloring is not in common use today. In the alloy coloring process, specific alloying elements create the apparent color by becoming dispersed into the anodizing bath and redepositing into the

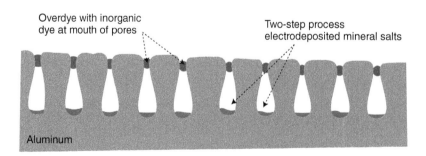

FIGURE 3.32 Process of overdyeing to enhance the appearance of the two-step process.

FIGURE 3.33 Two-step anodizing process with red overdye.
Source: Courtesy of AACRON.

pores of the oxide coating. In the 1950s and up until around the 1970s this technique was used to produce a matte gray anodized surface. The alloy A94043 was commonly used because of its silicon constituent. The silicon would come out into the electrolyte solution as undissolved particles and redeposit into the pores as they developed. This would interfere with light passing through the clear oxide and the surface would appear slightly opaque and matte gray. This process requires specific alloying control.

In 1923 the Bengough-Stuart process, using a 3% chromic acid solution as the electrolyte, was developed for protecting airplanes used in and around marine environments. This process produces a light- to dark-gray color by stepping up the voltage in stages. It is a time-consuming, expensive process still used today in the aerospace industry to provide an opaque, gray film with improved corrosion resistance.

It is not used in art or architecture, however. The film produced is thinner and softer than typical sulfuric acid processes.

Certain anodizing techniques involve the metal alloying constituents to develop color tones in the aluminum. The aluminum oxide that develops and thickens in the process will react with the alloying constituents and the electrolyte used to produce a color other than the light gray of a clear anodized layer. Some of those colors and the specific alloy are listed in Table 3.3. Not in current widespread use today, these techniques, called integral coloring, uses more energy and require more time than the two-stage process.

TABLE 3.3 Color achieved by reactions with alloying constituents.

Alloy	Sulfuric acid electrolyte	Oxalic acid electrolyte
1100	Silver	Dark yellow
3003	Beige	Pink-gray
4043	Gray-black	Green-black
5052	Green-yellow	Yellow
5083	Dark gray	Gray-brown
6061	Light yellow	Gray-yellow
6063	Silver	Gray-yellow

Integral Coloring

Integral coloring was a common process used in the 1960s and 1970s. It was introduced and developed by Alcoa and called Duranodic. They licensed others to perform the technique which developed color in a single anodizing step. This process involved organic acid baths where the organic acid would react with the dispersed aluminum oxide and redeposit into the pores. These colors have good weathering characteristics and ultraviolet resistance against fading, however, they would consume as much as three times the energy. The surface was hard and would color welds better than some of the processes used today.

The colors are limited to gray, black, bronze, and champagne tones. The process used a mixed electrolyte of organic and inorganic acids that reacted with the components in the alloy. To achieve a specific color, the Integral method was alloy critical. Certain colors could only be achieved with certain alloys containing specific alloying elements that would react with the organic acid in the electrolyte. The strict criteria for alloying made it more difficult to achieve consistency and made it more expensive than the two-stage electrolytic process more widely used today.

There are several trade names that used variations on the integral coloring process. A few of the more recognized names were:

Duranodic	Alcoa	Sulfuric acid with sulfosalicylic acid
Kalcolor	Kaiser aluminum	Sulfuric acid with sulfosalicylic acid
Alcanadox	Alcan	Oxalic acid

These have mostly been overshadowed by the two-stage electrolytic process of anodizing also known as the two-step process. Integral coloring requires greater energy use to produce the colors. Additionally, these finish processes are alloy sensitive. Small variations in the alloy in the sheet or plate would lead to color uniformity challenges. These finishes were only available by batch anodizing processes. They could not be coil anodized. Thicknesses of the anodized oxide would meet either the Class I, 18 μm thickness or the thinner, Class II oxide, 10–18 μm thickness.

Electrolytic Coloring: Two-Stage Process

The two-stage or two-step process is an anodizing system in wide use today. It serves an aesthetic purpose as well as producing a sound inert anodic coating on aluminum. For that reason, the majority of aluminum alloys that are colored using this method are the architectural alloys that fall into one of the four categories.

A 91xxx
A 93xxx
A 95xxx
A 96xxx

This process involves a second step where metal ions are deposited deep into the pores. The first step involves anodizing the aluminum part or sheet in the sulfuric acid electrolyte and developing a consistent porous surface. This is similar to the way aluminum has been anodized for decades. Degreasing, etching, and de-smutting processes are performed just as they are with other anodizing processes.

The second step, which was developed originally back in Japan by Asada and sometimes carries his name as the Asada Process, involves immersing the freshly anodized aluminum part into a bath containing a sulfuric acid electrolyte along with metal ions from metal salts dissolved in the bath. Instead of direct current, an alternating current is applied to the electrodes. This cycles the charge passing through the aluminum from positive to negative. This makes the aluminum submerged in the electrolyte at one cycle a cathode, or negative charge. The metal ions deposit onto the surface by cathodic deposition. The metal is deposited into the base of the pores because here the charge is most negative. During the other half of the cycle, the aluminum becomes an anode again and the porous oxide thickens further. This cycles back and forth for a predetermined amount of time.

Next the pores are sealed as before, using the hot immersion with additions of nickel acetate, locking the metal deposits into the pores as the aluminum hydroxide swells at the opening of each pore.

Several metal salts are used in the electrolyte to produce an array of colors. The resulting surface has good resistance to fade and superior weathering due to locking the metal ions deep into the pores versus on the surface of the pores. Additionally, since these are inorganic mineral-like deposits, they are not as susceptible to degradation from ultraviolet radiation as organic dye materials. The most common metallic salts are the tin and nickel salts. Metal ions that are used and the colors that are imparted are listed in Table 3.4.

TABLE 3.4 Metal salts and the color produced.

Metal ions and compounds	Color
Nickel	Bronze to black
Cobalt	Bronze to black
Copper	Maroon to black
Tin	Bronze to black
Ferric ammonium oxalate	Gold
Silver	Green
Gold	Violet
Selenium oxide (SeO_3)	Light gold
Tellurium oxide (TeO_3)	Bronze
Manganese oxide (MnO_4)	Bronze

Tin salts, such as stannous sulfide, are used to develop bronze tones. Nickel and tin salts are often combined in a ratio of 7:3 to produce different bronze tones (Figure 3.34).

The color achieved from this process is not derived from the metal salt color but from the behavior of light passing through the clear oxide and back to the viewer. Unlike color from inorganic and organic dye, this anodizing process achieves color from light-scattering behavior of the incident light passing through the oxide and reflecting off the base aluminum. The metal particles scatter the light and interference colors are perceived (Figure 3.35).

Because of the way color is achieved by light reflective changes, angle of view will have an effect on what the eye sees. Anodizing is not a paint or a coating but a conversion surface that is clear. The color in the electrolytic process or the two-step anodizing process is dependent on light energy and the interplay with the surface. It will not look the same in different lighting conditions because the energy of the light is different.

Overdyeing an aluminum surface that has received the two-step process prior to sealing the surface can produce remarkable, intense colors. These are subject to eventual fading, but the initial effect is deep color tones with a metallic appeal (Figure 3.36).

Removing the Anodized Surface

The aluminum oxide surface can be removed in the event there is an issue with color, or some surface staining has occurred. Immersing the anodized aluminum assembly into the etching bath will strip the oxide from the surface. The thickness of the part is affected and the removal of the problem that created the reason for stripping must be dealt with. If it is a color uniformity issue, then re-evaluation of the quality control procedures may be called for. Once stripped and etched,

Coloring by Means of Anodizing 143

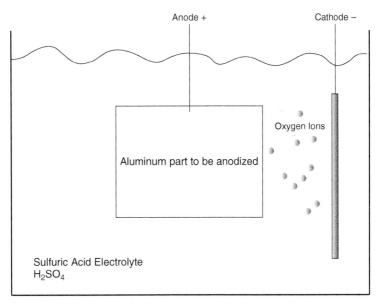

Two-step process: Step 1—Direct current application in standard sulfuric acid electrolyte

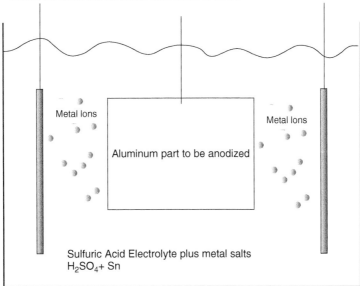

Two-step process: Step 2—Alternating current applied in sulfuric acid electrolyte along with the metal salt

FIGURE 3.34 Two-step process anodizing.

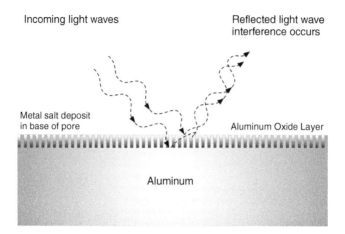

FIGURE 3.35 Light interference phenomena.

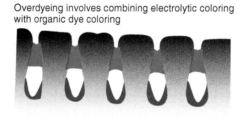

FIGURE 3.36 Organic overdyeing of an electrolytic colored aluminum.

the surface can be re-anodized; however, achieving an exact color match will be very difficult, if not impossible (Figure 3.37).

DESIGN CONSIDERATIONS

Variations in Color

With all natural surface material, you can expect slight color tone differences from one manufactured surface feature or series of features or panels to the adjoining group of panels. Aluminum is no exception. The control variables are too numerous, and these variables have their integrity defined by diverse processes in the production of the surface. Subtle differences can be manifest when perceiving the reflections off the surface. The presence of iron, silicon, or manganese tends to darken the initial clear anodic coating that develops. This can affect the color or tone exhibited even when a color is introduced. In a field of panel elements, for instance, several may be differing in color or

Design Considerations 145

FIGURE 3.37   Masking and etching designs on an anodized surface.

tone. In a series of extrusions, the reflective tone may appear as subtle differences in some, but not in others. This condition can change as you move around the surface or as lighting conditions change. It is normal and a natural occurrence not definable by known tolerances in processes.

The transparency of the anodized coating and the color and gloss that are perceived in the finished product depend on the anodized process used and the nature and purity of the aluminum alloy. Minute differences in alloying constituents on the surface can show color differences in both clear and colored anodized surfaces, even if the alloys and temper are the same. This is because processing at the mill may create conditions when anodized that can slightly alter the color that is seen.

The metal is cast in what is referred to as a lot or heat. This casting produces a number of ingots. Each ingot weighs approximately 13,000–14,000 kg (30,000 lbs.). The heat is one mixture of alloy within the tolerances of intermetallic elements added to the mix. Each of these ingots are similar in alloying consistency and this consistency is maintained in the subsequent rolling or extruding operations that follow.

Batch Anodizing

Batch anodizing involves complete immersion of the aluminum part into a fluid. Thus, hollow assemblies will float if the ends are closed. Slow draining assemblies and those with lap seams will trap acid. The acid will leach out and stain and etch the surface. For anodized parts, avoid hollow sections that will not drain or that will trap air and float. Thoroughly rinse the parts after removal from the acid electrolyte to ensure no excess fluid is trapped.

Since anodizing is an electrochemical process, the laws of electrical fields come into play. Small blind holes will only anodize to the depth equal to the diameter of the hole. Sharp corners will interfere with the electrical current in the electrolyte bath and generate areas that will not receive the current.

Designs that place thicker sections near thinner sections should allow for separate anodizing. If they are joined the current flow will be affected by the mass differences.

Welds will appear different than the surrounding metal. Regardless of which anodizing technique is used, the welds will appear lighter in appearance.

When anodizing, avoid steel inserts, steel bolts, and other steel fasteners. Be aware that batch anodizing requires the sheet, plate, or assembly to be hung from a rack. This requires a method of attachment, and often wires are used. These can leave a mark where they meet the assembly. If the electrode is attached to the part or assembly, this will also leave a mark. The mark is natural aluminum in appearance, appears like a burn or streak on the surface, and is more apparent on the darker anodized parts (Table 3.5).

Coil Anodizing

Another method in common usage for thin aluminum is coil anodizing. In coil anodizing the ribbon of metal is passed through the sequence of anodizing operations. The coil of aluminum is unwound and passed sequentially through a degreasing process, an etching tank, anodizing tank, coloring tank, and then a sealing tank, and then recoiled at the other end of the line. Available in thicknesses of 2 mm and less, this process offers excellent consistency and uniformity. It capitalizes on process speed and efficiencies. The drawbacks to coil anodized aluminum is the minimum quantity

TABLE 3.5 Batch anodize issues to overcome.

Issue	Problem	Remedy
Lap seams	Traps acid	Rinse thoroughly
Hollow/voids	Floats and holds acid	Drain holes, tilt to drain
Blind holes	Anodize partial depth	Full hole or eliminate
Sharp edges	Corner not as anodized	Round corner
Thick and thin joined	Anodize at different rates	Separate
Welds	Color difference	Match alloy or reduce weld
Steel inserts, fasteners	Corrosion and staining	All fasteners aluminum
Streaks from hanger	Unanodized region	Hanging system design

TABLE 3.6 Batch anodize versus coil anodize.

	Batch	Coil
Color consistency	Good	Excellent
Racking marks	Yes	No
Class I thickness	Yes	No
Class II thickness	Yes	No
Formed parts	Yes	No
Small quantity runs	Yes	No

requirements, an entire coil of metal is processed at one time, and the limitation on the anodizing thickness. The oxide thickness produced is a Class II and usually is only 10 μm in thickness. Any thicker and it lacks the flexibility in the metal oxide to be recoiled (Table 3.6 and Figure 3.38).

FIGURE 3.38 Grace Farms. Coil anodized clear. Designed by SANAA.

Other Design Considerations

The thermal conductivity of anodized films is approximately 10% of the base metal. The coefficient of thermal expansion is only 20% of the base metal. If the anodized aluminum surface reaches elevated temperatures exceeding 80°C (176°F), then tiny, spiderweb-like cracks will appear on the surface as the underlying metal expands more than the anodized aluminum.

Anodized films are ceramic-like in nature and this makes them lack flexibility. The anodized film is inflexible and will show small cracks if subjected to elongation of more than 0.5%. A bend in an anodic film will crack along its length. This crack is only in the oxide layer, but it tends to pull away from the expanding underlying metal.

The anodized aluminum surface can be pierced with a drill or a hole can be punched, but there will be small cracks that could alter the appearance around the pierce mark.

Welding anodized aluminum requires the removal of the anodized film. The heat from welding will crack the coating near the weld as the underlying metal expands.

Anodizing is an electrochemical treatment performed on aluminum of any shape or form. Anodizing is one of the most common surface treatments used on aluminum in art and architecture.

Paint Coatings on Aluminum

There are several challenges in painting aluminum, which is inherently more difficult to work with than steel. The aluminum surface is highly reactive to oxygen and forms a thin aluminum oxide spontaneously on exposure to air. Where the oxide is stripped by chemical action or mechanical action, it must be immediately coated or when the aluminum surface is exposed to the air the oxide will rapidly reform. The aluminum surface requires preparation for good paint adhesion. If the pretreatment is not performed correctly, the paint will flake off the surface (Figure 3.39).

Anodizing is an excellent pretreatment for aluminum when it comes to paint adhesion because it is nonreactive and stable. This inert surface, with its porous microscopic landscape offers the ability for paints to "key" into it. For this to work effectively, however, the anodized surface should not be sealed. Once sealed properly, the anodized surface will not readily accept paint.

Chromic acid anodizing or chromate pretreatments, which used to be the norm in paint preparation for aluminum, are now carefully regulated due to problems with disposing of the hazardous chromium solutions. Phosphoric acid and boric acid anodizing can work as well to provide a pretreatment to painting and improve paint adhesion. There are new treatments for aluminum that provide good paint adhesion. These should be discussed with the paint applicator (Figure 3.40).

Design Considerations 149

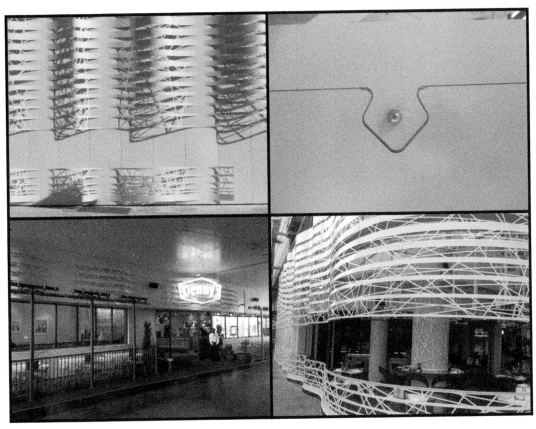

FIGURE 3.39 Painted aluminum surface.
Designed by James Wines Architect.

Coil Coating

There are various types of paint used to coat aluminum. The types and performance expectations of these paint coatings are beyond the scope of this book. It is important, however, that aluminum be properly prepared in order to enable the paint to adhere.

Prefinished aluminum sheet metal materials are readily available in many colors and paint types. These are superior coatings applied via coil coating processes that over the past several decades have been highly perfected. Simply look around at all the facades of flat composite panels and you recognize the consistency and durability of the finish on aluminum sheet material.

These finishes are applied on coils of aluminum in a highly control process. These processes, by their nature, require a sizable quantity, a consistent and predictable aluminum surface and a relatively thin material, generally less than 2 mm in thickness. The finish paint is applied in a

FIGURE 3.40 *The Chrysalis at Merriweather Park*, painted aluminum sheet, by Mark Fornes.

programmed process to a long ribbon of metal. The process etches the aluminum surface, applies a primer to the surface and applies a finish coating to the surface followed by heating the aluminum in an oven, all while the metal ribbon is moving. The process does not allow stopping and placement of a few sheets to achieve a color. It is a coil process and requires a sizeable quantity of metal to be processed at a given time (Figure 3.41).

The process of coil coating is well refined and controlled today, which makes it the highest quality of paint finish available on aluminum sheet material. The difficulty is that it has thickness limitations. Coil coating cannot be performed on thick plate larger than 3mm.

Custom Painting of Aluminum

For requirements that call for special colors of limited quantity, painting thicker aluminum or painting formed and fabricated aluminum, there are several methods that are commonly used. Each of these methods has its limitations and challenges. Some of the paint coatings are excellent, but the process is not nearly as refined or controlled as coil coating and expectations should be adjusted to account for the increase in variables.

FIGURE 3.41 Coil-coated aluminum used on the Bloomberg Center. Designed by Morphosis.

The aluminum surface with its quintessential oxide is not the best for accepting paint. Some simple paint processes will work but most art and architectural exposures require the aluminum surface to undergo pretreatment if they are to be painted.

Chemical or mechanical means are the most common pretreatments. Hazardous chromate pretreatments are being eliminated for many processes due to the toxic nature of the hexavalent chromium that is used. Hexavalent chromium, Cr^{+6}, is considered to be a carcinogenic compound and toxic to organic life. Compounds that carry chromium in the +6 oxidation state can pose health concerns mainly to those applying the coatings or working with them. Chromate pretreatments have been used on aluminum for decades and remain the best pretreatment found for the metal. The formula for the chromate pretreatment is:

$$6Cr(OH)_3 \cdot H_2CrO_4 \cdot 4Al_2O_3 \cdot 8H_2O$$

Phosphate pretreatments and silage pretreatments are chemical processes that are commonly used in the pretreatment of aluminum surfaces for receiving paint. There are several other proprietary methods that involve reducing or eliminating hexavalent chromium and instead use trivalent chromium, which is considered less hazardous. Chromium phosphates are coatings used on canned goods made of aluminum and is indicated by the pale green color found on the interior of aluminum cans.

Mechanical treatments are also being used more regularly in the pretreatment of aluminum. After degreasing and just before the coating process, the surface of the metal is blasted with clean sand, shot, or glass beads, just before the coating process.

Paint Type	Application	Benefits
Halogenated resins—fluorocarbon finishes	Coil, wet, and powder	30–40-year service life. Durable and UV resistant
Acrylics	Powder and wet	Good UV resistance
Epoxy	Powder and wet	Chemical resistant, poor weathering
Urethane	Powder and wet	Durable, limited exterior colorfastness
Polyesters—super durable polyesters	Powder and wet	Limited exterior durability as compared to fluorocarbon finishes

Powder Coating

Electrostatic painting, also known as powder coating, is a process that uses powdered, dry beads of paint to coat aluminum, and for many applications works very well to provide a thick protective coating. A positive charge is applied to the aluminum. Tiny beads of dry, negatively charged particles of various polymers are sprayed toward the aluminum part. The charged particles find exposed areas of aluminum and stick to it electrostatically. Polymers often used are polyesters, polyurethane, epoxy, and acrylics. The aluminum, once coated with the small polymer beads adhering to the surface, goes into an oven where the temperature is sufficient to melt the beads and fuse them together.

There is no solvent involved as with liquid coatings and waste is minimized as excess paint particles can be collected and recycled. Vertical surfaces can be coated with powder just as efficiently as horizontal surfaces without issues of runs or pooling. The coatings are thick and smooth, and the thicker coatings, approximately $50\,\mu m$, provide better smoothness. Thin coatings applied with powder tend to show "orange peeling," a pebbly surface texture.

There are three steps to the process of painting aluminum with powder: preparation of the aluminum surface, application of the powder, and curing in an oven. Preparation of the aluminum surface is critical for good adhesion of the coating. The surface must be dry and free of oils and grease. Any surface contamination will alter the charge and insulate the surface.

Coating process involves passing the powder through a gun, sometimes referred to as a corona gun. This provides the powder with a charge and as it leaves the gun it seeks out any surface with the opposite charge—in this case, the charged metal. The particles stick evenly to the surface because of the electrostatic attraction.

Once the piece is coated, it passes through an oven with sufficient temperature to melt the small polymer particles and create a thin coating of cross-linked polymer over the surface. The curing temperature is $200°C$.

Powder coatings are difficult to repair in the field if they are damaged. Powder coatings by their nature cannot be field finishes, so a wet coating must be formulated to match. This leads to differences in appearances.

The ability to recycle and reuse excess powder is one of the main benefits of powder coatings, but it also leads to a potential problem in many powder application facilities. The powder coating is applied in a booth and a negative pressure within the booth collects the excess powder and sends it back through the electrostatic gun. However, this negative pressure can also pick up dirt and moisture, and even oils that may be in the vicinity.

Another issue with powder coatings and all electrostatically applied coatings is the Faraday cage effect that occurs around perforations and blind holes. The Faraday cage effect repels the charged particles as they enter holes in the surface. While the edges get coated, the charged particles do not coat the sides or bottom of the hole.

Wet Coating

Large aluminum forms can be painted with what are often referred to as wet systems. These involve solvents and can be either oven cured or air dry. Both cases involve the application of a coating with a solvent carrier, where the solvent evaporates and leaves the resin and pigment to remain. They can be aided by electrostatic attraction, similar to powder. The paint is mixed and applied via spray gun or roller to the pretreated aluminum surface. The cleaning, degreasing, and pretreatment of the metal surface is critical with the wet paint application processes as well. Cleaning and degreasing are similar to other paint processes. The surface must be thoroughly cleaned of greases, oils, and oxides. Pretreatment is a very critical step for the wet coating applications. Chromate pretreatments are the workhorse when it comes to aluminum pretreatments but restrictions on the use of hexavalent chromium are pushing the industry to other, safer treatments.

Wet systems are composed of solvent carriers, resins, or polymers, pigments for color, and other additives to give special affects to the finish. The polymers used are polyesters, polyurethanes, acrylics, and epoxies (Figure 3.42).

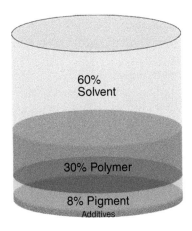

FIGURE 3.42 Makeup of Wet System.

154 Chapter 3 Surface Finishing

FIGURE 3.43 Wet applied over aluminum plate, by Jan Hendrix. Student Center in Qatar University.

Wet painting presents several advantages over other systems. One is the use of metallics in the paints to produce pearlescent and iridescent colors with the introduction of metal flake or mica flake into the paint. Mica is a mineral that is nonconductive. Metal flake is usually aluminum powder. Both are mixed into the paint and applied to the surface. Mica is used more often when electrostatic paint application is used because mica does not align itself in the direction of the charge. Mica provides a beautiful pearlescent to the reflectivity. The aluminum powder increases the reflectivity of the painted surface (Figure 3.43).

Wet paints have the advantage of being easily mixed to create colors. For example, adding red pigment to yellow pigment gives an orange color. However, with powder coating, mixing a red powder with a yellow powder gives you red and yellow powder, not the color orange. See Table 3.7.

Patinas

Aluminum can be patinated. It is not considered as a good candidate for patina development because of the tenacity of the aluminum oxide. The oxide can be altered to create beautiful dark tones, but not

TABLE 3.7 Comparison of wet and powder coating.

	Wet coating	Powder coating
Small lot processing	Yes, can mix small lots.	Yes, if stock colors
Solvent	Yes	No
Mixing to create colors	Yes	No; preformulated
Blending colors	Yes	No
VOC (volatile organic compounds)	Yes	No
Electrical current needed	Not necessary	Required
Oven	Large or Air dry	Small
Ventilation	Yes	No
Shrinkage	Yes	No
Recycle excess	No	Yes
Film thickness	Thin films to thick films	Thick films
Metallic coatings	Yes	No
Repairable	Yes, easy to match	No, difficult to match

without difficulty. The tones ranges from deep black to mottled grays to dark brown tones. There are several proprietary processes available to darken aluminum. They all begin with a very clean surface. If the oxide layer can be stripped, the reaction of the surface and the stability of the coating in most environments is adequate to good without an addition of a clear coating (Figure 3.44).

The corrosion resistance of aluminum makes patination very difficult. Most mineral forms developed on the aluminum surface tend to the gray to black appearance. Darkish brown tones can also be generated but these lack stability.

The dark blacks and dark grayish black mottled surfaces are stable. The black can be very porous, and dirt and dust will cling to the surface when used outdoors. These will appear lighter against the deep black background and the contrast may be less appealing. Due to this porosity, the deep black surfaces are difficult to clean and maintain. Sealing the surface will help, but because of the intense deep black color the surface will hold heat, which can cause the coating to decay prematurely.

Ferric chloride can etch the surface and impart gray tones. Depending on the alloy, the tones can have a rusty or even slightly green tint. You will need to apply a clear seal coat over the surface to maintain it.

Texturing the surface can generate effects of mottling when patina solutions are applied. One can darken the surface, then bring back highlights to enhance the appearance. These are proprietary cold patina processes (Figure 3.45).

156 Chapter 3 Surface Finishing

FIGURE 3.44 Rosenthal Museum of Contemporary Art. Blackened aluminum. Designed by Zaha Hadid.

FIGURE 3.45 University of Southern Indiana Performing Arts Center. Mottled darkened aluminum. Designed by Malcolm Holtzman.

Patination on aluminum does not offer much choice to the designer and artist. Anodizing is the most effective way to achieve color depth and variety (Figure 3.46).

Chemical Milled Surface

By placing a resist over the surface of aluminum sheet, plate, or casting, a decorative surface can be created by immersion in an agitated bath of a chemical etchant. The process dates back to the days of the suit of armor when highly decorative etching was inscribed into the iron plates. For aluminum it was developed for the aerospace industry back in the 1950s. Today it is a common technique for removing weight in airplane manufacture. An unbroken resist is applied to the aluminum. The resist could be cutout and applied, painted onto the surface, or created by means of photosensitive resists.

Once the layer is applied, the aluminum is placed in a tank of etchant, usually a strong alkaline mixture, such as sodium hydroxide along with other etching fluids. Where the aluminum is exposed and not covered by the resist, the etching solution attacks the aluminum and dissolves a layer at a time.

The resulting finish is a matte surface. The depth is determined by the amount of time and concentration of the chemistry. As the metal is removed, a smooth contour draft is created rather than an abrupt 90 angle. The limitations are the size of the tank and depth of the etch. The etching process can completely perforate the sheet, but because of the phenomenon of draft angle, the edges of the perforated hole will be razor sharp (Figure 3.47).

FIGURE 3.46 Darkened light sconces, by Ron Fischer

After chemical milling, the surface can be polished to create highlights, or it can be anodized. All aluminum alloys can be chemically milled.

Machined Surface

Aluminum can be machined using all forms of milling machines. Aluminum is a perfect material for machining. Table mills, 5-axis mills, end mills, and even routers work well in cutting and shaping the aluminum surface. The mill should be operated at high cutting speeds with or without lubricants, high rake angles, and high relief angles. The alloys with high silicon are more abrasive and harder tools are needed. Very intricate surfaces can be created rather quickly with aluminum. After machining, the surface can be anodized, polished, or left with the machining lines.

The alloys that machine best are the A91xxx, A92xxx, and the A97xxx alloys. Additionally, the harder tempers machine better. Aluminum has a tendency to wrap around the tooling as it is

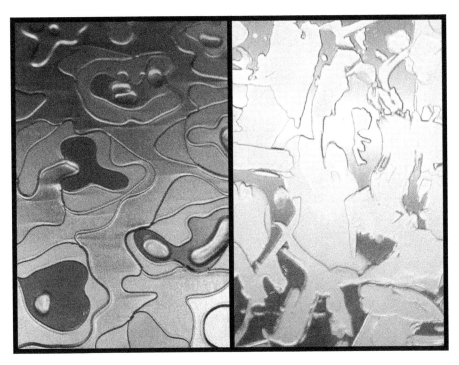

FIGURE 3.47   Chemical etched surfaces, courtesy of Metalix Desighn.

removed. This can slow down the process when stopping to peel the sharp aluminum coil from the tool (Figure 3.48).

Finishes on Extrusions

Aluminum extrusions can be finished in manners similar to those for other wrought forms of aluminum. Anodizing is a common finishing technique. Extrusions are batch anodized. That is, they are assembled one at a time onto an electrically conductive rack that holds several lengths of extruded aluminum. These undergo the same process as plate or sheet. Limitations depend on the size of the anodizing tanks. Most tanks are limited in size because of the cost of heating the various baths used in the process.

Clear and color anodizing are all available for extrusions. Rare occasions may call for removing scratches and glass bead blasting the aluminum surface, but typically the extrusion is anodized directly with the extrusion surface only, similar to sheets and plates.

On hollow forms, the extrusion must be open to allow the fluids to enter and the processing must allow for time to drain the fluids from the interior of the tubular form. The insides of the extrusion receive only a slight amount of treatment as compared to the outer surface.

Painting extrusions is also a common finishing process. Again, the limitations are the size of the paint booth used to apply the paint. The aluminum needs to be pretreated to receive the paint which if a chromate is used, requires a tank large enough to accommodate. For some processes, such as

FIGURE 3.48 Machined aluminum surface.

powder coatings there are size limitations in the ovens needed to fuse the small paint beads. The interior of the extrusion is left unfinished in the tubular forms.

Finishes on Castings

Aluminum castings can be finished. Some alloys will anodize better than others and the casting process used will have a definite effect on the end appearance of the finish product.

Aluminum castings will need to have the initial finish applied mechanically. The removal of the gates and risers, performing all welding, and then finishing the surface by grinding and blasting to arrive at a consistent appearance will be required before anodizing or paint application.

Anodizing large cast sections may struggle with achieving consistent results. The variations that will appear on a large cast surface are created from the grain development on the cast surface. Developing a good anodized surface on casting requires careful finishing and cleaning. It is advised to work with the foundry and the anodizing facility to arrive at an understanding of the finish expectations.

Painting a cast part will also entail special attention to the finish surface. Surface porosity is the most common issue confronted, particularly with the sand cast method. Die casting will produce the best surface for painting or anodizing, but this is a costlier process and deals with smaller parts.

CHAPTER 4

The Aluminum Surface Finish: Meeting Expectations

Change the Environment; do not try to change the man.

—Buckminster Fuller

We see aluminum every day in the products we use, the transportation system we travel on, and the buildings we work in. It is as ubiquitous a material as wood or ceramic, yet when we consider it for large surfaces, surfaces made from numerous elements or custom forms, we sometimes struggle with the overall impression and our expectations.

Is the color silver, gray, or white? Does the surface reflect a monolithic or patchwork appearance? Is the surface wavy or smooth? Why does the surface look different with the angle of viewing or the time of day? These questions of appearance define the subjective nature of our expectations of the aluminum surface.

Aluminum surfaces in art and architecture are provided in one of three surface finishes:

Natural. The color of the metal itself with no enhancement other than mechanical abrasion or surface imparted from the rolls or dies used to produce the form.

Anodized. Conversion coating of a thickened oxide that may be with or without added color dyes or effects.

Painted. Organic coatings, both clear and pigmented, applied over the surface to impart color.

Inorganic coatings are not common as a finish coating for aluminum. Inorganic coatings would include porcelain enamel, plating, and coatings of chromate compounds. Plating aluminum is difficult because of the adherent oxide and not often performed in art and architectural uses of the metal. Porcelain coating requires high temperature to fuse the porcelain to the aluminum and this produces

162 Chapter 4 The Aluminum Surface Finish: Meeting Expectations

thermal dimensional changes that can be a challenge. The differences in the coefficient of thermal expansion of the two materials is significant and can lead to cracking as the aluminum cools in contrast to the glass-like porcelain, which has a significant different coefficient of thermal expansion.

Material	Coefficient of thermal expansion $\times 10^{-6}/°C$
Porcelain	4
Aluminum	22

THE NATURAL FINISH

By natural finish, we mean the finish that aluminum displays without any induced oxide or coating. This color is what is produced when alumina is refined, cast into ingots, and rolled into plates or sheets, extruded through dies to produce various shapes, or cast into molds. The finish is what is imparted to the surface during those processes or mechanically applied after those processes, such as blast texture on sand cast aluminum or mechanically applied directional and non-directional finishes.

The challenge with aluminum often coincides with how to specify and achieve a consistent surface appearance or, in actuality, how to interpret what a consistent finish will be. The natural surface of aluminum can be beautiful and display a predictable and even appearance when produced from a single heat casting and if the surface finishing procedure is maintained throughout the process (see Figure 4.1).

FIGURE 4.1 Exceptional refined aluminum mill finish by AMAG.

The Natural Finish   163

The natural aluminum finish will weather and darken slightly over time. It should be avoided in high traffic areas because it can and will scratch easily. If scratched, there is no easy way to return it back to the original surface without treating the entire area. This is no different with any metal. Removing the scratch requires removing part of the surface and this brightens this area. Mechanical finishes applied to aluminum will darken quickly as they collect grime, fingerprints, and dirt. Cleaning the natural surface when a mechanical finish has been applied can be difficult. Because of this most natural finish applications either remain as the mill provided surface and color and used in areas of low traffic or embossed or they are coated or anodized (see Figure 4.2).

FIGURE 4.2   Desks made of mill finish aluminum plate, designed by Foster Partners.

164　Chapter 4　The Aluminum Surface Finish: Meeting Expectations

As with any material, if you do not occasionally clean the surface, you can expect it to collect dirt and grime deposited from the surrounding environment. Inside or outside, metal surfaces will collect airborne substances. Aluminum is no exception. But this occurs over time and is highly dependent on the coarseness of the surface, the rainfall patterns, geometry and the quality of the surrounding atmosphere.

Certain geometries and designs limit whether a surface can be anodized or pretreated sufficiently to hold paint. A poorly adhered paint surface can be far worse in appearance than on oxidized natural aluminum surface.

Assemblies that require a significant amount of welding may limit the options to the natural finish appearance. Large, thick sections may preclude painting as an option because of limitations of oven size or the limitations of pretreatments to get the paint to adhere. Many light poles around the world are made of aluminum and have nothing more than a light brushed finish. They have been shown to hold up sufficiently well. Art work, created from large aluminum tube forms, exposed in an urban environment for more than 20 years with only natural rains to clean them, appear as if they were recently installed (see Figure 4.3).

FIGURE 4.3　Bartle hall sculptures, after 20 years' exposure, by Ron Fischer.

There are many instances where natural finished aluminum has performed sufficiently well over years of exposure with little more than the occasional rain to clean the surfaces. If salt exposure and dissimilar metal contact is minimal, you can expect natural, uncoated aluminum to perform well.

Aluminum has specific characteristics influenced by intrinsic properties of its surface, mechanical behavior, and microstructure. As you approach a surface made out of metal elements, the color and edges define the geometric form. With the exception of a painted aluminum surface where the pigment in the paint plays a more significant role in appearance, surfaces made from aluminum have a strong metallic quality that can appear patchwork as the individual elements reflect the light in different ways. Diffuse aluminum surfaces will show the patchwork aspect more than a mirror polished surface. This is due to the way matte surfaces tend to show the real color tone of the material and less reflection of the surroundings. Matte surfaces, at the micro level, have ridges and valleys or craters with overlapping microscopic rims. Anodized aluminum normally has a diffuse texture.

These rough surfaces influence the optical properties of a metal by trapping some of the light and reflecting it at different angles back to the viewer. Unlike a mirror surface that reflects the surroundings and their color, a roughened aluminum surface scatters the light and the eye captures more of the color of the surface. Aluminum reflects a significant portion of the visible spectrum, which gives it a white appearance when the surface is roughened. The newly installed aluminum surface without paint or anodic coating, looks bright and white-gray. Figure 4.4 shows the wavelength reflected from the surface of pure aluminum. Aluminum reflects over 90% of the wavelengths

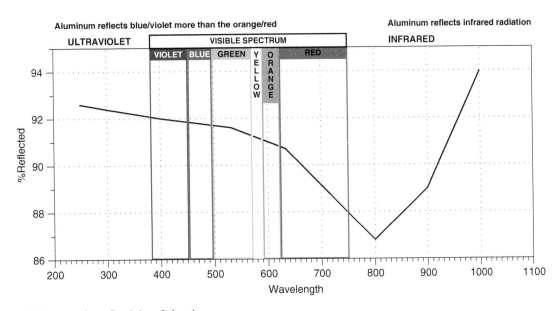

FIGURE 4.4 The reflectivity of aluminum.

FIGURE 4.5   Handrails made of aluminum pipe and water jet cut fins, before and after 20 years of exposure.

associated with purple, blue, green, yellow, and orange. The red wavelength is slightly less but still more than 85%. This reflectivity makes aluminum a choice for mirrors and produces a white appearance, tending slightly toward the blue end of the spectrum. As aluminum weathers, the surface darkens and loses some of its gloss. The darkening comes from pollutants and hydroxides that form on the surface. In the satin finish, aluminum does not have the same luster as stainless steel; it has a softer reflective quality. Because of its surface clarity and color, it can be mistaken for stainless steel. Side by side, it has a whiter color tone than the bluish tone of stainless steel (see Figure 4.6).

As with any material, if you do not occasionally clean the surface, you can expect it to collect dirt and grime deposited from the surrounding environment. Inside or outside, metal surfaces will collect airborne substances. Aluminum is no exception.

FIGURE 4.6 Comparisons of non-directional satin aluminum with non-directional satin finish stainless steel. Note that the aluminum reflection appears whiter. This is due to the way aluminum reflects nearly 90% of the visible spectrum while stainless steel reflects 60%.

THE ANODIC FINISH

In the majority of applications of aluminum, the metal is coated or anodized. Coatings are typically organic paints applied via powder application, wet spray application, or coil coating. These finishes are performed on clean milled surfaces that have been pretreated to receive the paint. There are clear coatings for interior applications that can be applied directly to satin finishes on aluminum or over glass bead surfaces. The appearance of such coated surfaces depends on the integrity of the coating itself and how it adheres to a surface, not so much on the metal itself. The metal could be any alloy or any other metal, for that matter.

The anodized surface, however, maintains the metallic appearance of aluminum. Anodizing is an extension of the aluminum and not specifically a coating, but a transition oxide film formed by electrochemical means. As one approaches an anodized surface, the aluminum will reflect slight variations from segment to segment, panel to panel, or extrusion to extrusion. The angle of view can also play a role in the appearance of anodized aluminum elements that make up a surface. Some viewpoints will show no difference in color while others may show a relatively large contrasting appearance (see Figure 4.7).

These variations in apparent color and reflectivity between anodized aluminum elements are due to several factors that are not in the control of the producer, manufacturer, or anodizing company. It is the natural character of the material. Measurements taken of the coating thickness should show the oxide film to be very close from element to element. This is one method used to help determine if the oxide is of similar thickness between elements and is one of the quantifiable standards. However, identical oxide thicknesses between anodized elements may still show appearance differences. Minute subtleties in the grains can create a differential in the reflectivity of light. Slight alloying element differences between heats may show contrast in color between two elements. Whether the surface is clear anodized or electrolytically color anodized, the differences can, and often will, occur.

FIGURE 4.7 Color differences are visible in the same alloy but from a different heat. All panels were batch anodized at the same time.

The industry has struggled with this for decades, as designers have turned to the beauty of anodized aluminum as a surface material. The aluminum producers of sheet metal and extrusions are constantly refining their processes to improve the predictability of the anodic surface, but in the end, the accumulated tolerances of the processes of production through the processes of installation make the exact matching of anodic elements of a surface impossible. Coil anodized material comes the closest particularly when color is introduced. However, two surfaces, even if they are from the same sheet, will appear differently if the angle of reflectivity is different. The beauty of the metal is the reflective, metallic quality, and angular differences in reflectivity can create appearance differences (see Figure 4.8).

The anodic surface is more stable than the nonanodized aluminum surface when it comes to atmospheric pollutants. The process of anodizing aluminum first etches the surface in a caustic bath of sodium hydroxide. This gives a matte reflective quality. This etching process leaves the aluminum surface very clean because it takes off a few molecules of the outer layer. It also can concentrate the intermetallic elements that make up the aluminum alloy. These intermetallic elements interplay with light at the quantum level and can slightly alter the appearance of the aluminum surface.

For instance, alloys of aluminum containing a maximum level of copper can have a slight champagne appearance when anodized. The tiniest amounts of copper on the surface will move the light reflectivity just a touch more to the orange end of the visible spectrum.

These subtle variations in tone are not something that can be repaired or adjusted. It is the natural beauty derived from taking an element composed of small grains and minute surface irregularities that formed during the intense heat and pressure of the original metal production.

FIGURE 4.8 When viewed at an angle, color differences appear.

170   Chapter 4   The Aluminum Surface Finish: Meeting Expectations

We often think of industrial processes as being under tight controls. Many manufacturing processes must fall under tolerances imposed by the interface of one part to another, tolerances of mechanical production and expansion and contraction from temperature fluctuations. The finish on the surface of aluminum is influenced by roll wear, lubrication, wear on abrasive belts or pads, minute marks on the die used to extrude the part, and slight variations in the fluids and throwing power within the anodizing tank. These are variables that over the years have been refined and halved and adjusted until standards of acceptance are achieved. Like a Zeno Paradox,[1] there is no possible way tolerances can reach the point of perfection (see Figure 4.9).

How do you convey inconsistencies and acceptable tolerances in a world where perfection and perfect surface appearance reside only within the confines of thought? For example, a person's

FIGURE 4.9   Wyly theater, designed by Rex.

---

[1] The Zeno Paradox of Achilles and the tortoise relates how, whenever Achilles arrives at where the tortoise has been, he still has some distance to go before he can even reach the tortoise. This can be an analogy for those moments when you approach perfection, but still have some distance to go.

TABLE 4.1 Variables to contend with at different levels in the processing of aluminum.

Inherent in metal	Mill process	Fabrication process	Environmental use
Alloying elements	Alloying elements	Quality of forming dies	Exposure
Grain structure	Mechanical properties	Quality assurance processes	Cleaning regimen
Stiffness	Internal stress	Exposure to other metals	Light exposure
Reflectivity of metal	Quantity limits	Anodizing color	Lifespan expectations
Oxidation tendency	Smoothness of dies and rolls	Anodizing thickness	Tolerance of structure
Oxide color	Production tolerances	Paint type	Surface coarseness
Recycle quantity in alloy	Metal thickness	Paint process	Expectations of the end user
Cost of production	Hot or cold rolled surface	Finish type	Reflection of surroundings
	Cost	Cost	

perception of "what is achievable" may be derived not from the reality of the process but from ungrounded expectations. Small samples of finishes can serve the designer as guidelines, but they often fail to give the bigger picture presented by the final overall project. This is inherent in all natural uncoated metals and the oxides induced on the surface, either naturally or artificially. See Table 4.1.

There are forces pulling against one another that require balancing, but the end result is not in the center, where the ideal resides, but somewhere off to the side. Understanding the interplay of these forces and how to adjust them to arrive at an achievable and acceptable appearance is what must occur. This is not to suggest that the designer should accept mediocrity or sloppiness, but instead to recommend that the designer anticipate the interplay beforehand and use it to achieve something remarkable in the end. Buy-in and understanding must be determined by all parties in the process.

THE POINT OF RANGE SAMPLES

It is thought by many in the design and construction industry that acquiring a range sample is sufficient to arrive at the absolutes that a color or tone will display. The only time this is truly possible is at the time of actual production of the metal into its final form. By then a lot of effort, energy, and cost has occurred. Rejection would be expensive, schedules would be up-ended, and no one wants to take this risk.

If all projects could begin from a stocking source where all the metal needed for an application already resides, then one could selects the specific metal in the exact finish. But this is never the case except for small projects, and even then challenges can develop. The metal surface may appear different when viewed in the actual setting.

Range samples produced ahead of production are highly subjective. In reality they represent the degree of difference that could be allowed, not the actual, absolute boundaries of color and tone ranges. They cannot be the true representation. They are of a different time and the finish was developed under a different set of physical conditions. The process and procedure may be identical, but the chemistry and thermodynamics can never be. The industry demands objective criteria for metal surface appearance, but it is in reality wishful thinking. Why? It is a matter of cost, timing, thermodynamics, and quantum behavior of the molecules making up the surface.

Metals are produced in a furnace and cast as alloys, which are nothing more than mixtures of metals and other elements. The mixtures are massive in weight and are subject to small variations in distribution of the alloying elements and thermodynamic behavior of the cooling of masses of metal. Once the metal is cast, it is first scalped to remove impurities that have floated to the top as the mass of metal cooled. Then it is rolled under heat and pressure to produce sheet and plate or extruded through heated dies. The rolls impart a particular character to the surface, as do the extrusion dies, as well. At this point, a lot of energy, cost, and time has been expended and at this point only, when the metal has been produced, is it possible to get the expected range in color and tone. You can't tell the precise grain of a piece of wood until you cut the tree down and saw it into planks, and it is a similar case with metal. You cannot send it back at this point any more than you could put a tree back together. Too much expense and energy has occurred.

The quantum part has to do with how the surface is viewed. Metals absorb light only to a few molecules on the surface. The free electrons on the surface are excited and the light is re-emitted, giving a metallic luster. This is what conveys the metallic aspect of a metal, such as anodized or natural aluminum. This is also what creates variations in color tone as various elements are assembled to form an overall surface.

The texture of the surface plays a part in final appearance as well. The texture captures the light and sends it back to the viewer. If the surface is mirror polished, then intensive, hot spots, will mask changes in the base color, but if the texture is matte, the diffuse reflection will enhance minor variations. These variations will show under different light energies depending on what time of day it is or what the lighting source is. Sometimes a surface made of various elements will have one element that dominates over the adjacent elements. This could be lighter or darker. Often, when the light energy changes, this difference disappears and can even flip, making the dominant element appear darker or lighter than the surrounding elements (see Figure 4.10).

This should not mean that we are stuck with whatever the mill happens to produce. There are unacceptable conditions that can and do arise in the production of the metal. These are due to procedures or processes that fell out of tolerance during production. Qualified producers of the material will identify these before they are presented to the user. Unfortunately, some can result in

FIGURE 4.10 Image of different color tones in sun and in shade.

delays in delivery as the replacement material is reconfigured, recast, rolled, and extruded. Some of the surface appearance issues arrive once the piece is anodized. Streaks due to inconsistencies in the alloying can show as darker or lighter lines. These are particularly troublesome because the metal has been cast, rolled, or extruded and is in the final stages of finishing. They are not visible or indicative in the sheet or extrusion until the oxide is enhanced. See Table 4.2.

COLOR

Metallic luster on anodized aluminum, in particular dye-anodized aluminum, can be striking. The dye enters the pores of the anodic coating produced in the anodizing process. The top of the pore is sealed. This clear film, along with the clear oxide, create depth and richness along with the color.

TABLE 4.2 Procedures to consider to improve success with aluminum surfaces.

Consider the following to ensure success:

- Develop a quality assurance plan that involves the supplier of the metal and the fabricator of the product.
- Specify the correct alloy. This will be an outcome of working with the metal supplier and fabricator to achieve the end result.
- Specify the correct temper. This will be an outcome of working with the metal supplier and fabricator to achieve the end result. It is dependent on how the aluminum will be processed.
- Require a mill certificate verification of alloying constituents and mechanical properties of the metal produced at the mill. If the quality assurance plan involves the supplier, often you can tighten some of the production tolerances.
- For wrought products, require leveling and elimination of stretch lines, Lüder lines, coil breaks, and chatter.
- Extrusion die samples should be pushed and reviewed for surface quality acceptance. These should be of same alloy and temper as the project requires. This will give a good representation of what the surface of project metal will possess. These should be of sufficient size, no less than 500 mm.
- Obtain representative samples of similar alloy and temper aluminum before a finish is applied for surface review. This will provide an indication of the micro-texture that the project metal will possess. It will not be the final metal but should have a similar surface quality. Obtain at minimum three, 250 mm by 250 mm.
- Obtain representative samples of similar alloy and temper aluminum with similar final finish applied. Obtain at minimum three, 250 mm by 250 mm.

    These will not be the quintessential range samples but should provide a good representation of what one can expect when the final product is produced.
- Establish and agree to, with the supplier and fabricator, what criteria will be used for rejection or acceptance. The criteria must be as objective as practical and signed off on and agreed to by all.
- Once the metal is produced, obtain final coupons of the aluminum form before finish and after finish. Compare these to the representative samples. This should be for all wrought forms used on the project, extrusion, sheet or plate, or tubing or bar. The samples should be similar in size for an adequate analysis.

The richness of the color defines metallic luster. Unfortunately, though, this effect is short-lived in a material performance context. All dye anodized will eventually fade as the chemical bonds break down from ultraviolet light exposure. The aluminum is reflective, and this helps extend the color, but dulling and breakdown of the color will eventually occur (see Figure 4.11).

Depending on the color formulation and the way the color is imparted into the pores of the aluminum, the environment the surface will be exposed to and how often it is cleaned will determine the extent to which a color will fade and decay. Interior surfaces, where ultraviolet light is minimal, will keep their colors longer. Fluorescent lighting will emit small amounts of ultraviolet

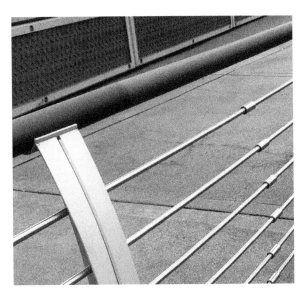

FIGURE 4.11 Image of color degradation of anodized aluminum over 18 years exposure. Bronze handrail color change on part of the rail.

radiation, as will incandescent lights. LED lighting produces no ultraviolet radiation. Sunlight exposure, however, takes a toll on the aesthetic lifespan of dyes used to produce some of the colors in anodized aluminum. The ultraviolet light wave contains sufficient energy to break the carbon bonds in organic dyes. Once this occurs the color will fade.

The inorganic dyes and the electrolytic two-step anodizing process afford ultraviolet radiation resistance to a far greater degree than the organic dye anodizing. This resistance to fade is called "light fastness" and is a property of the pigment or dye to resist change over time and exposure.

In the case of organic and inorganic dyes, there are various rating systems for light fastness as defined by the Blue Wool Scale and by the American Society for Testing and Materials (ASTM). See Table 4.3.

TABLE 4.3 Light fastness rating.

Light fastness property	Blue wool scale	ASTM
Very poor	1	V
Poor	2–3	IV
Fair	4–5	III
Good	6	II
Excellent	7–8	I

For architectural and art use where a dye is considered, a good to excellent rating should be used. The inorganic dyes will be the better choice for many instances because they will have a better light fastness rating. The inorganic nature of the mineral salts will not decay like the organic dyes. Inorganic dyes achieve their color from metal salts. These salt particles are larger than the pigment molecules and offer a variety of colors. They are not as bright as the organic dye pigments, but they are more resistant to fading and are usually less expensive.

The electrolytic two-step process uses inorganic metallic salts deposited deep in the pores of the metal to produce color by means of light interference. The available colors are limited, as compared to inorganic and organic dyes; however, the resistance to fade gives them an advantage over the dye processes. See Table 4.4.

Expectations of perfect color consistency is a challenge that the industry has worked to address. Color is not the sole criteria to create a unique appearance. Gloss and texture also play a significant role in establishing the consistency of an aluminum surface. Color can be measured with a colorimeter, but color can be affected by gloss. Gloss can be measured with a gloss meter, and there are various types of gloss meters in use. They measure the reflection off a tiny segment of a surface by projecting a beam of light at a particular angle. For highly reflective surfaces, an 80° gloss meter is used. For matte textures, a 60° gloss meter is used. On satin, directional finishes, the gloss measurements will be different when taken perpendicular to the grain versus taking the measurements parallel to the grain. The readings are relative to the meter used. When measuring gloss, you must take a number of readings across a surface and arrive at a mean. Remember that you are taking a representative sampling of the surface. Otherwise the noise in the readings will give false comparisons.

Profilometers can provide a measure of roughness but they do not provide sufficient information for establishing how a particular roughness of a surface will affect the play of light off that surface. Optical profilers provide much more information about the nature of the surface texture.

Some optical profiler devices can provide color analysis, gloss, and information about the texture of the surface. Optical profilers provide very detailed information about a surface but differences in measurements can be subject to misinterpretation. You must correlate the results with an acceptable range that the human eye perceives. A surface can give different readings but still appear the same to the human eye. Additionally, optical profile readings are difficult to obtain in situ. These devices are laboratory devices.

TABLE 4.4 Expected time until fade becomes apparent.

Finish type	Interior (low UV)	Exterior UV exposure	Exterior UV and tropical exposure
Organic dye	10–15 years	1 year or less	—
Inorganic dye	30 years plus	10–12 years	2–3 years
Electrolytic	30 years plus	15–20 years	8–12 years

In considering the color and color comparison, it is important to arrive at an acceptable and practical method of evaluation. An effort should be made to consider how the surface will be viewed in the final installed position. For an interior application, the type of lighting that will strike the surface should be considered as well as the angle of view (see Figure 4.12).

One accepted practice is to view the surface from approximately 3 m away. The grazing angle of the light source, time of day, and angle of sun, if the sun is the light source, should be consistent. This technique works well for many surfaces as a means of establishing some boundaries within the analysis of whether a surface is correct or not. Blemishes in the surface and other minor conditions disappear at this distance. Color and gloss as viewed by the human eye can still be subjective, but the gloss and color measurements provide some objective criteria to arrive at an acceptable condition when reviewing the surface intended for a project. Viewing this way enables a designer to eliminate small issues in the surface finish.

In practice, this method can fall short, in particular on surfaces with a diffuse texture. As the surface is viewed from a distance between 30 and 50 m, distortions begin to reveal themselves as the reflected light vectors show dark and light tones created from very slight surface plane changes.

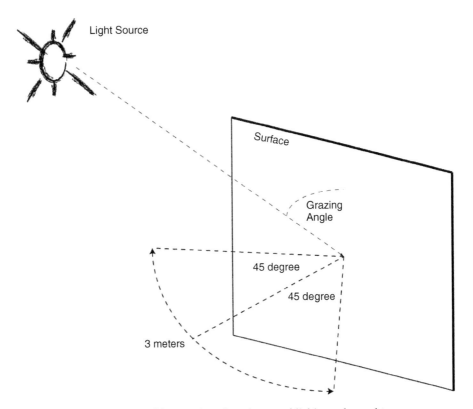

FIGURE 4.12 Diagram of viewing positions and angles when establishing color and tone.

This phenomenon is similar to what occurs when looking at light reflecting off water. Up close, the surface may look smooth, even glasslike, but as you view the water from a greater distance, undulations can begin to appear. Digital images can capture this better than the human eye. What can look like differences of several millimeters, but when measured, are only a fraction. Even colors can have different appearances as distance changes. What may appear to the eye as a good match can look different when viewed at a distance or through a camera.

This creates a dilemma because the criteria that are chosen in a specific spatial condition cannot be duplicated as measurable criteria for the manufacturer when viewing the material in a plant setting. The lighting and the angle of view will be significantly different in a factory setting than when installed in the final setting.

Range samples produced ahead of production are highly subjective. In reality they represent the degree of difference that could be allowed, not the actual, absolute boundaries of color and tone ranges. They cannot be the true representation. They are of a different time and the finish was developed under a different set of physical conditions.

FLATNESS

Proper packaging for transport of the parts through fabrication and on to the project site is of critical importance as well. Inducing stress in the assembled parts or the flat sheet forms or plates can influence how the finish product will appear (see Figure 4.13).

Careful allowance for the softness and lower strength of the metal will produce the results needed for success. This should be a consideration for design as well as performance.

Flatness expectations for aluminum surfaces made from sheet or plate are heavily dependent on allowances for thermal expansion and contraction, leveling and thickness. Aluminum has a low melting point metal and this correlates to a more significant coefficient of thermal expansion.[2]

For aluminum, the coefficient of thermal expansion is a range between 21×10^{-6} m/m°C to approximately 24×10^{-6} m/m°C.[3] The range varies due to alloying constituents. For all practical purposes in art and architecture the difference is minimal, thus the value of 23×10^{-6} m/m°C is often utilized. This value means for any given length of aluminum, as the temperature changes from an initial level to a different level, either warmer or colder, the length of the aluminum will change. All materials will change in size and volume as the temperature changes.

If, for example, a 3050 mm in length element of aluminum is being shaped and formed in a plant that is operating at around 10°C and the aluminum part is an extruded linear form installed in an exterior environment where the metal either heats up or cools down as the temperature or solar

---

[2]James, J.D., Spittle, J.A., Brown, S.G.R., and Evans, R.W. (2000). *A Review of Measurement Techniques for the Thermal Expansion Coefficient of Metals and Alloys at Elevated Temperatures*. Institute of Physics Publishing.
[3]Engineering ToolBox (2008). *Linear Thermal Expansion*. [online] Available at https://www.engineeringtoolbox.com/linear-thermal-expansion-d_1379.html [accessed August 17, 2018].

FIGURE 4.13 Importance of proper packaging.

conditions change, then the metal can be expected to change in dimension approximately as the graph predicts. If the temperature of the metal reaches 40°C, the 3050 mm length will grow more than 2 mm over this temperature difference.

To determine how much thermal expansion to expect or to design for, use the following formula with the table of coefficient of thermal expansions for similar alloy types (see Figure 4.14).

$$\Delta L = L_i \times \partial (t_f - t_i)$$

ΔL = Change in length expected
L_i = Initial length of part
∂ = Coefficient of thermal expansion
t_f = Maximum design temperature
t_i = Initial design temperature

$$\Delta L = 3050 \times 0.000023 (40 - 10)$$

$$\Delta L = 2.10 \text{ mm}$$

The coefficient of linear expansion of a metal must be taken into account when installed in conditions that experience significant temperature changes. Aluminum has a high thermal conductivity, which helps keep the temperature distributed across the metal's form. Aluminum objects

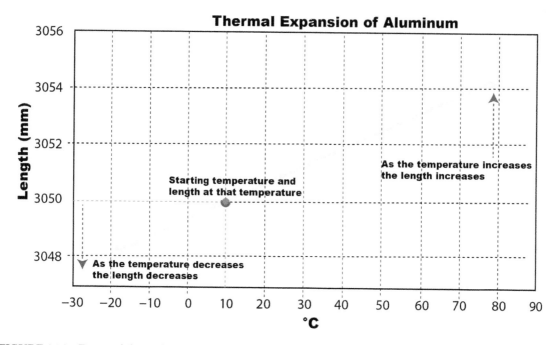

FIGURE 4.14 Expected thermal expansion and contraction of a given length of aluminum.

tend to be slightly cooler unless they are darkened. Aluminum is a good reflector of infrared. This is why it is used as foils to wrap foods, keeping the heat reflected back into the food.

When designing and constructing with aluminum, the change in the dimension of the metal must be taken into consideration. If the material is bound or thermal conditions are not taken into consideration, it may manifest as distortion on the surface, commonly referred to as "oil canning" or "pillowing" or, in the worst cases of overstress, "connections" and "cracking."

Using slotted connections that allow for slippage without binding the metal or eroding the metal are recommended. It is important, however, that the aluminum be fixed at some point, preferably an end, and all thermal expansion and contraction is to or from this point. A point in the middle can also be considered but thermal movement will be in all directions away from the point of fixation. Depending on the stiffness of the shape, this will induce stress as the aluminum expands against and has to push against friction and gravity forces. Warping of the surface can occur or premature wear at the point of contact may occur (see Figure 4.15).

Aluminum sheet or plate should be leveled to remove internal stresses that have developed and accumulated during strain hardening and temper rolling. The result is a distribution of the stresses across the sheet or plate in such a way that the stresses are negated. There are several methods that level the material. Often referred to as stretcher leveling or roller leveling, the sheet or plate is passed through a series of upper and lower small rollers that alternate moving the metal up and down. The rolls can be adjusted to subject the sheet to more deformation than other regions in the

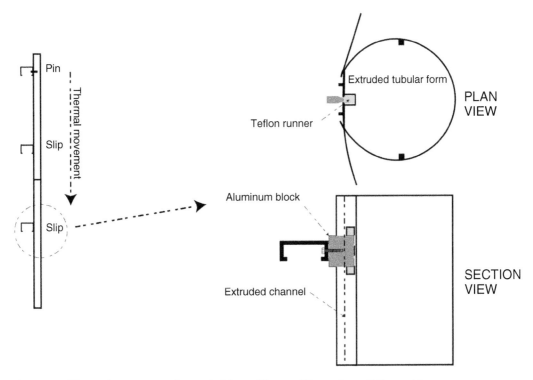

FIGURE 4.15 Thermal movement of extruded form. Slip condition using a Teflon guide.

sheet in the case of differential stress. This removes the residual stress caused by coiling the metal. This stress, if not removed, can show as curvature along the length of the sheet or laterally across the sheet. For heavy plate, the metal is often stretched at the mill to flatten the plate. The result is a smooth, flat surface with minimal internal stress.

Extrusions and other linear sections such as bar and tube are stretched and flattened by passing through a set of rolls or tugging on the ends. This imparts slight elongation into the part and removes differential stress, which can cause shaping and curvature.

For flat surfaces, there is no substitute for thickness. Stiffeners applied to the reverse side, adhered backing, and deformation ribs pressed or rolled into the sheet can all help to a point, but for aluminum thickness is important for flatness. Fusion studs used to attach stiffeners onto the reverse side of a thin sheet or specially formulated adhesive bonds are common means of adding section and stiffness to thin aluminum plates. Caution is needed to ensure there is no visible telegraphing to the face side. Again, thickness plays an important role in not translating effects to the visible surface. Minimum thickness for the use of these stiffening techniques falls at 2 mm. Thinner material will telegraph the stiffeners through to the face surface, and they will become apparent on the face side. It takes skill and quality assurance procedures to achieve repeatable results (see Figure 4.16).

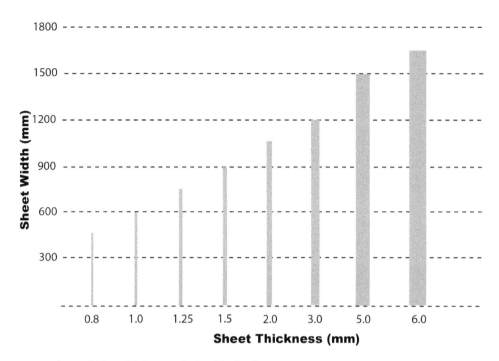

FIGURE 4.16 Sheet width to thickness relationship for flatness concerns.

The matte reflective character of the anodized aluminum surface will partially mute surface distortions. The lower yield strength of aluminum alloys used in art and architecture will limit the extent aluminum can span between supports. Deflections will also be greater due to the low modulus of elasticity of aluminum. You will need to engineer the connections with strength and stress concentrations in mind. Fusion stud welds weaken the aluminum at the point of fusion and large flat surfaces need to be engineered to distribute the loads. With aluminum it is critical to eliminate strain concentrations at narrow shear bands where shear strain can develop as the metal undergoes deformation (see Figure 4.17).

CLEANING THE SURFACE OVER TIME

If the metal surface is not maintained and cleaned after a period of exposure, it will eventually show dirt and grime. The mill aluminum surface that is not coated with paint nor anodized to give it an enhanced oxide will darken as the natural oxide thickens and combines with various components of the atmosphere. Many industrial plants built in the 1960s and 1970s, were clad in corrugated aluminum panels. The alloys used were the A93xxx alloys. They have darkened with age but still perform very well. The clips and fasteners holding them in place were galvanized steel or cadmium-plated steel fasteners.

Cleaning the Surface Over Time 183

FIGURE 4.17 Naval Memorial Canopy, designed by BNIM Architects: 4 mm thick natural aluminum finish. Large panels allowed to drape under their weight.

The aluminum surfaces used on these plant walls are still in good shape. The siding sheet was relatively thin but sufficiently strong to span from steel purlin to steel purlin. The oxide has combined with other components of the atmosphere—in particular, the industrial atmosphere—and formed dark carbides in some places.

How much cleaning and how often and what methods are acceptable for the cost depend on the environment that the surface is exposed to, the coarseness of the metal finish, the ease of access to perform cleaning, and, not least, the expectations of the end user.

Constraints to Consider on Cleaning the Surface
- Expectations of the end user
- Environment
- Coarseness of surface
- Accessibility
- Regularity of cleaning
- Method of cleaning

When cleaning aluminum surfaces, consider neutral pH cleaners, those with a pH in the 4–8 range. Alkaline cleaners or acidic cleaners can harm the aluminum surface if not neutralized quickly. Aluminum surfaces are more sensitive to mildly alkaline cleaners and solutions than to mildly acidic solutions. Tiny breaks in the oxide film of natural finish aluminum open the base metal up for pitting corrosion attack when continuous exposure to alkaline cleaners occurs. Adding sodium silicate can help inhibit the attack from stronger alkaline cleaners.

Cleaning of metals should take into consideration the safety and health of the people who will be performing the cleaning and the surroundings where the cleaning will occur. Ensuring the safety and health of the people doing the cleaning should at a minimum involve protection for eyes and skin and a source of clean, fresh water to be used to dilute a spill or for an emergency wash-down. When using chemicals, be certain to obtain the Safety Data Sheet (SDS) on the product and follow the instructions.

Safely disposing of the spent cleaning material and protecting the adjacent materials should be part of the cleaning plan.

There are three basic types of aluminum surfaces when considering methods of cleaning.

1. Natural finish
2. Anodized finish
3. Paint finish

For any cleaning procedure, start with the simplest and safest method. Deionized water and neutral pH detergents such as dish soap should be considered first, to see if they are sufficiently effective in removing contaminants from the surface. For aluminum, it is recommended to expose the surface to substances with pH that falls within the range of 4–8. Deionized water may start out at a neutral pH of 7 but, when exposed to carbon dioxide in the air, the pH will drop below 7. Most dish soaps have a pH slightly above 7.

The cleaning process is used to displace the adhered dirt and grime, convert it to a solute, and remove it from the surface. Deionized water, if it is available, will grab the ions in the soot on the surface and allow them to be washed off the surface. Deionized water is much more effective than normal tap water when it comes to attracting the minerals in dirt and soot. Dish soaps are polarized. The soap molecule has an end that is highly attracted to water while the other end is attracted to oils and grease. When used, the soap molecule attaches to the oils on the surface and the water rinses the combination away from the surface.

More aggressive cleaners can be used on painted and anodized surfaces. However, avoiding strong alkali cleaners is advised. If they are used, then have a method of neutralizing the base and flushing the surface with fresh water to remove as much of the alkali as possible.

Alcohols such as isopropyl alcohol have a pH of 5.5 and will aid in the removal of heavier grease and fingerprints. Follow this with a clean water rinse to remove the residues that are left behind. On dye-anodized colors, you should test a section in advance to ensure you are not disturbing the dye and displacing it. Removing fingerprints and light soils with 99% isopropyl works well without leaving a residue or film on the aluminum surface.

There are many biodegradable, industrial cleaners available that have a neutral pH. They will work well to displace the dirt and grease. Warming the solutions usually improves the performance (see Figure 4.18).

If you use cleaners such as glass cleaners, ammonia, bleach, or borax cleaners, these are all alkali cleaners with pH in the range from 10 to as high as 13. If they are allowed to remain on the

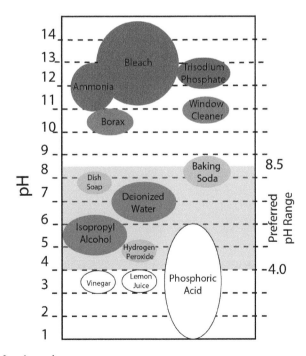

FIGURE 4.18 The pH of various cleaners.

aluminum, they will etch the surface and possibly damage the appearance. Anodized aluminum is subject to etching as well, if the cleaning fluids remain on the surface. Anodized aluminum is used on curtain wall supports. Because of this, it is recommended that the cleaners used to clean the glass are thoroughly removed and not allowed to remain on the aluminum horizontal surfaces supporting the glass. Vinegar and lemon juice are pH 3 and are acidic. They can help remove stains from other metals, such as rust from corroding steel, but they must be neutralized and thoroughly rinsed from the aluminum surface after use or they also can etch the glass.

Organic solvents will be necessary to remove adhesives, gums, waxes, and lacquers. Organic solvents such as acetone, mineral spirits, xylene, toluene, stoddard solvent, and naphtha will not affect the aluminum surface and will work to remove organic substances. Exercise safety practices and disposal care in using solvents.

Heavy oxides, water stains, and blackish stains on natural aluminum surfaces cannot be removed with soaps or solvents. Acid treatments will be required. The use of acids to remove the oxides and stains can be delicate as well as hazardous. There are cleaners available at various automotive stores that are suggested for use on aluminum wheels. The cleaners that work effectively have fluoride acids that should be handled with extreme care. There are other aluminum gel type cleaners that will remove minor tarnish on aluminum, but deep dark stains are difficult to clean without removing the oxide layer. Mechanical methods of sanding or blasting may be needed. When using these acidic treatments on aluminum, once the cleaning is completed, the surfaces should be neutralized and any residues and acids thoroughly rinsed from the surface.

For anodized surfaces, as with painted surfaces, cleaning should be with mild detergent and water. The use of solvents will not affect the anodized surfaces. Organic dye color anodize can be damaged if high temperatures coupled with detergents are used. High temperatures can open the pores.

The more polished the aluminum surface, the easier to clean. Surfaces with coarse textures will hold dirt and grime, and oxidation can occur in the valleys of the coarse surface. Cracks, gaps, and overlapping seams can hold moisture, deicing salts, and cleaning solutions. Allowing them to remain can create small corrosion cells that develop into thickened oxides.

CHAPTER 5

Designing with the Available Forms of Aluminum

Don't take it all too seriously. If you want to live your life in a creative way, as an artist, you have to not look back too much. You have to be willing to take whatever you've done and whoever you were and throw them away.

—Steve Jobs

Aluminum is one of the most versatile metals known. It possesses unique properties that afford the designer and fabricator a tremendous variety of options and combinations. Initially aluminum was made into casting using conventional sand molds to create lightweight yet strong forms. The Wright brothers made the engine for their first successful powered flight out of an aluminum cast block that was later milled to fit cast iron pistons. Some of the first artistic expressions of aluminum were made from castings. The metal required less energy and if the porosity was held in check, the surface was sound (Figure 5.1).

It did not take long for the establishment of other production processes for use in industry, furniture, and art. Cast blocks of aluminum were heated and rolled into plates and sheets that could be turned into corrugated paneling for lightweight cladding. The ductility of aluminum sheet allowed it to be hammered into various shapes by forging, spun into disks and bowl forms, stamped into panels, and extruded into linear forms of intricate cross-sections.

The dawn of aluminum coincided with the development of the automobile and the development of the airplane. Without aluminum, the development and growth of both inventions would have been hindered.

188 Chapter 5   Designing with the Available Forms of Aluminum

FIGURE 5.1   Custom cast aluminum for a wall in Sweden.

Attributes of Aluminum
- Lightweight
- Clarity of appearance
- Light reflection quality
- Corrosion resistant
- Malleable
- Ease of Extrusion
- Good strength
- Anodizing
- Ease of milling
- Heat strengthening
- Adjustable alloy selection
- Low melting point
- Castable
- Recycle ability

BASIC FORMS OF ALUMINUM

The wrought forms of aluminum consist of wire, foil, coil, sheet, plate, and extrusion forms. These are the forms that are produced at a mill from large controlled castings and reshaped and modified under pressure into useful forms for industry. The initial casting made at the mill is specific to the intended wrought form. Castings for plate, sheet, and foil are made from large blocks of rectangular cross-section, while those for extrusion and wire are made from large cylindrical blocks with a circular cross-section.

Wrought Forms

- Foil
- Sheet
- Plate
- Extrusion
- Wire
- Pipe
- Tube

The other major category is cast aluminum. Aluminum is ideal for all casting processes because of its low melting point. There are several types of casting in common use to make aluminum products and forms. Sand, permanent mold, and die casting are the most common techniques used for the casting of aluminum.

Cast Forms

- Sand casting
- Permanent mold casting
- Die casting
- Investment casting
- Centrifugal

Further, there is another form of aluminum that falls outside of these two categories but finds use in art and architecture.

Foamed Aluminum

Not all alloys are available in each form. If it is a matter of matching color or finish of a design across forms—for example, from extrusion to plate or casting to sheet—there are ways to arrive at this, but it should be understood that the finish, grain structure and exact chemical makeup of the alloy will be slightly different from one form to the next. Subsequent anodizing process may intensify these subtle differences.

IT BEGINS AT THE ALUMINUM MILL: THE HEAT

For the intent of appearance and quality, it all starts at the mill. At the aluminum mill, ingots are produced from alumina refinement and recycled scrap. These have predetermined alloying constituents that fall within an industry established range. Each ingot will have substantially the same alloying makeup; however, ingot to ingot, the composition will vary slightly even when the alloy number is identical.

In what is known as the "heat," these ingots are established and cast into specific shape depending on the intended use. There are extrusion ingots, sheet ingots, and primary foundry ingots. Each has a specific shape defined by the process they are intended for.

The mill will provide on request a "Mill Certificate," which certifies that the material produced conforms to the specifications outlined in the order and the adherence to established industry specifications. The mill certificate contains the information shown in Figure 5.2.

The language may be slightly different from one mill to the next, but the basic information is the same. All aluminum producers test the material to insure it conforms to standards set out by the industry or specific requirements from the customer.

For large orders, obtaining a mill certificate should be a standard practice. This ensures that the Mill met the quality standards and specifications for the production of the alloy and it ensures that the mechanical requirements were verified by a qualified metallurgist.

Even small orders will have an associated mill certificate on hand to verify that the alloy makeup and mechanical properties adhere to industry criteria. These certificates should be on hand and are traceable back to the original source of production.

SHEET AND PLATE

Sheet and plate ingots are large blocks of rectangular cross-section. The top of this large block is removed in a process called scalping. Most of the impurities and oxides that form rise to the top when the block is cast, and this portion is cut from the slab and removed and recycled to capture any aluminum. Edges are often sawed as well to square the sides up to fit a particular rolling mill operation.

This large block is heated in a furnace and passed through rolls at the rolling mill that put intensive pressure on the hot block of aluminum. This widens the block as it becomes thinner and longer. The hot aluminum passes several times, incrementally becoming wider, thinner, and longer. Eventually the desired thickness is achieved, and the thick block is now a thin plate of aluminum. If plate is the end product, then the edges are trimmed, and the ribbon of metal is cut to lengths and skidded for heat treatment or surface treatment. If the thickness is less than 9 mm, the ribbon of metal may be recoiled. This rough coil of plate material is set up to be cold rolled, in which it passes back and forth between pressure rolls that impart smoothness to the surface and establish a level of strain hardening while the metal is further thinned. To adjust the strain hardening the coils are set in an annealing oven to soften the material to a desired temper, or it can be placed back on the cold rolling pressure rolls to be thinned further (Figure 5.3).

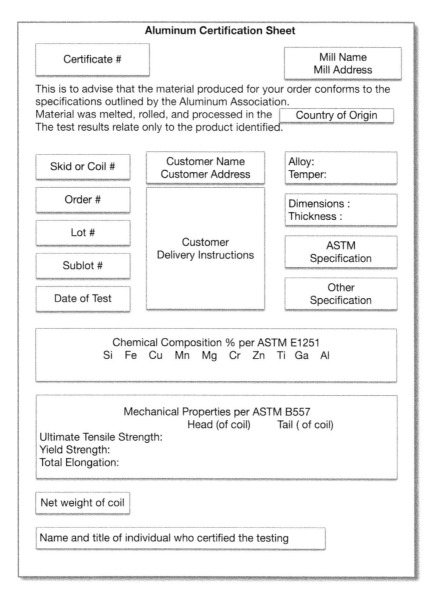

FIGURE 5.2   Typical information on a Mill Certificate.

SUPPLY CONSTRAINTS

In the marketplace, the supply of aluminum is dictated first by the Mill. The Mill has limitations of minimum quantities when it performs its initial castings. Many projects involving aluminum fall under these minimums. A Mill requires a set minimum order size for a particular alloy and a particular temper. This minimum often is several thousand kilos of aluminum.

FIGURE 5.3 Plates of aluminum, wrapped, labeled, and protected in a plant.

To overcome this, there are warehouses that stock various popular alloys in tempers and set dimensions. There still may be minimum quantities subject to skidding charges, but availability of the more popular alloys and tempers, many of which are the architectural alloys, are stocked.

Another limitation will be on the thickness of the aluminum in stock, particularly with plates. Thinner sheet can be stored in coil where it is leveled, decoiled, and skidded to order. Plate thicknesses are not as readily available, and it is recommended to check with supply houses for width and length inventory.

Architecture and art play a large role in the use of aluminum, and the quality of the surface is paramount. On large thicknesses and with certain forms—such as large diameter tubing or pipe, rod, and bar—the surface may not be in pristine shape. Nearly always, the alloy number and temper are printed on the surface. Some marring and staining as well as scratches may be on aluminum coming from a stock house. This will lead to added finishing costs. Anodizing processes will not remove these surface finish issues.

Supply Constraints
- Minimum quantities
- Availability
- Stocked sizes and thicknesses, alloys, and tempers
- Surface quality

Tube, wire, rod, and some extruded forms that are popular dimensions and alloys are stocked at supply warehouses similar to sheet and plate. There is not a wide range of alloy and tempers to choose from. For instance, if you are purchasing a half dozen 6-m tubes of 150 mm diameter and 1 mm wall thickness, you may have to alter your design to accommodate what supply is available.

Custom extruded shapes are a different matter. In these cases you will be subject to minimum quantities. First, you will need to pay the cost for the die and die design. Then there is usually a minimum quantity of aluminum that will be pushed through the die, regardless of how much you need. More often there are limit ranges where you must take the full quantity pushed, even if you only need a portion of that quantity.

When you are ordering from the Mill or from a large extrusion producer, you may be subject to a condition where they can supply 10% more or 10% less than your order. This condition can be extremely costly to the user. It forces you to order 10% more than you need to ensure you get the full amount. In reality you could end up with as much as 20% more than the project requires. A good relationship with a supplier is critical.

PLATE

Aluminum plate is defined as a rectangular form with a minimum thickness of 6 mm (0.25 in.). The edges are either sheared or saw-cut and squared. Plate can be finished to the final thickness in hot rolling processes or cold rolling processes.

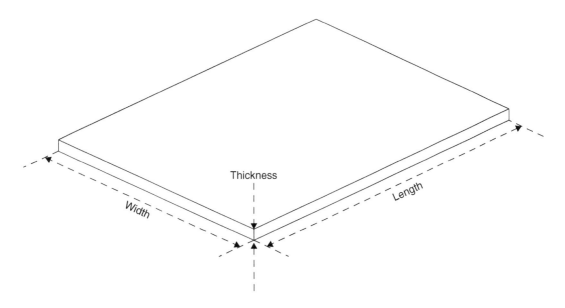

For plates there are several widths and thickness that are more commonly produced (Table 5.1). The alloy and temper limit some of the potentially available plate sizes. Length is a function of handling and flatness. If the plate length exceeds 3 m, special skidding will be required. For large plates, handling and logistics play a larger role in determining the cost. If the quantity needed is not a mill run, then you are subject to available stock sizes in both alloy and dimensions.

TABLE 5.1 Typical widths available for plate.

Available widths	
Inches	mm
36.5	927
48.5	1232
60.5	1537
72.0	1829
84.0	2134
96.0	2438

Plate thicknesses are considered to be 6 mm (0.25 in.) or heavier. These are hot rolled and then cold passed to improve the surface and to adjust the mechanical properties by cold working. You can order hot rolled, finished plate, but the surface is rougher and not as precise as thinner aluminum. The roughness is due to the grain quality and size. On thick plate the "as fabricated" surface has a distinctive grain that is larger than that on sheet forms. There are numerous alloys available in the following plate thicknesses (Table 5.2).

TABLE 5.2 Nominal thicknesses for plate aluminum.

Plate thicknesses	
Inches	mm
0.25	6.35
0.31	8.00
0.38	9.50
0.50	12.7
0.68	17.1
0.75	19.0
0.88	22.2
1.00	25.4
1.25	31.8
1.50	38.1
1.75	44.5
2.00	50.8
2.50	63.5
3.00	76.2
4.00	101.6

Aluminum plate is different from steel plate. As with steel, aluminum plate is hot rolled but the subsequent pass through cold rolls imparts smoothness to the surface. The grain may appear larger, but the surface is flat and has fewer inclusions than a similar stainless steel or steel plate.

The Mill will cut the plate to specific lengths. Variable lengths and blank sizing will be performed at a secondary fabrication facility, usually not performed at the mill producer (Figure 5.4).

Aluminum plate used in architectural and art is typically in either A95xxx series or A96xxx series. These alloys give good workability and are corrosion resistant. Within the A95xxx series, the following specific alloys are often considered for projects where plate is considered.

A95005	H32 or H34 tempers
A95052	H32 tempers
A95086	H32 or H34 tempers

The tempers are induced by cold working the plate then stabilizing the stain hardening by natural or artificial aging. The hardness is the quarter hard or half hard levels.

The A95xxx series plates are more susceptible to a visual surface imperfection known as Lüder lines or Lüder bands. These appear on the surface of sheets as stripes. The stripes typically appear as parallel lines that run diagonally across the width of the sheet. They can also appear as flowing bands. Chapter 8 has a more in-depth discussion of Lüder lines.

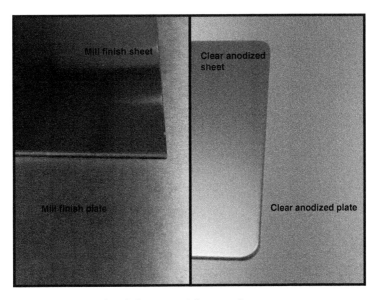

FIGURE 5.4 Grain of sheet versus grain of plate material comparison.

The alloy A96061 is another excellent plate material used in art and architecture. It is less prone to develop Lüder lines and has good strength and corrosion resistance.

A96061   T4, T6, and T651 tempers

The surface flatness and surface quality are very good, and a variety of plate dimensions are available. The temper stands for solution heat treatments where the aluminum is held at a predetermined temperature of approximately 532°C (990°F). The T4 temper indicates that the aluminum was naturally aged to stabilize the mechanical properties, while the T6 temper means that the aluminum was artificially aged to achieve maximum strength. The T651 temper means the plate underwent stretching to relieve the internal stresses that develop from the rapid quenching of the heated aluminum. Stretching is usually 1–3% in the direction of the length.

When ordering plate material, it should be understood that the Mill will print the alloy and temper onto the surface along with the specific standard used to produce the plate. This printing can be removed with solvents. If instructed, the Mill will not print on the plates at the time of producing the plate material (Figure 5.5).

The surface finish on plate aluminum can be very good but still lacks the smoothness obtained from sheet or coil. Because it is plate, there is a limited amount of passes through cold reducing and thus the grains are not as tight and elongated. There can be small surface imperfections in plate material that will influence the quality of final finishing steps (Figure 5.6).

FIGURE 5.5 Mill ink print on plate surface.

FIGURE 5.6 Small pits and inconsistencies in an aluminum plate.

COIL AND SHEET

The initial aluminum melt or heat is always to a specific alloy. Certain ingots are cast to be converted into coil and, from there, decoiled and sheared to sheet lengths. This initial large block of aluminum is scalped to remove the impurities that rose to the top as the metal recrystallized into the solid state. Edges of this slab of aluminum are usually sawed to even them out for subsequent rolling and reducing.

This slab is positioned onto a rolling stand and heated, passed through pressure rolls that squeeze and reduce the thickness, eventually reaching a predesignated thickness. This reduction in thickness while hot sets out the grain direction in the metal sheet. Further reduction is done cold as the cooled metal is passed under more polished rolls that impart a smooth surface on the aluminum and strain harden the grains by cold working the metal back and forth between sets of cold rolls. This is referred to as temper rolling. Once a certain cold rolled temper is achieved the coil is set into a heat treatment bell to stabilize the temper by artificial aging or a thermal treatment to relax the metal by annealing to a different temper level. All of this is performed in a regular, prescribed process with the goal to arrive at the desired mechanical properties.

The cold work on polished rolls imparts a surface finish to the metal. The coil of aluminum is leveled and cut to specific sheet lengths. It may be slit into thinner coils for specific processes or it may remain in the large coil form or recoiled into smaller coil sizes to meet specific manufacturing needs.

The criteria for sheet aluminum, like plate, is of rectangular form and rectangular in cross-section. The thickness is less than 6 mm (0.25 in.), but not less than 0.20 mm (0.006 in.). In the United States the minimum thickness for sheet is 0.15 mm; less than that is considered foil. All sheet and coil material achieve their final finish in the cold rolling process.

Coil sizes are defined by width and weight. Arbor size is the opening in the middle of the coil and corresponds to the mechanical spool that holds the coil for decoiling into blank sheet sizes, roll forming, or turned into smaller coils for further use. The arbors are available in diameters as small as 76 mm (3 in.) for special narrow slit coils and diameters as large as 610 mm (24 in.). The weight of the coil produced at the mill are usually very large. Coils can be further processed to reduce the size and thus the weight for specific handling needs. Roll forming shapes, for instance, often require smaller coils and coil weights (Figure 5.7).

All sheets come from coils of the particular alloy and temper. The aluminum is strain hardened by a process of cold working the metal. This involves passing the long ribbon of metal back and forth between sets of cold rolls. Once a particular temper is achieved by cold working, the aluminum is recoiled. The heat treatment processes also occur while the aluminum is in the coil form.

Coils of aluminum are provided in various widths (Table 5.3). It is critical to verify availability from a particular mill source. Aluminum is available in several widths depending on the mill, the alloy, and the thickness of the metal. Other widths can be accommodated by slitting the coils to narrower widths. This is usually performed at a coil processing facility.

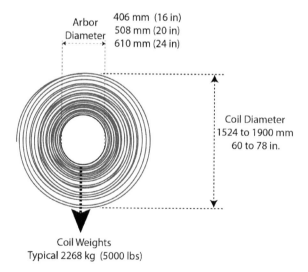

FIGURE 5.7 Typical range of aluminum coil dimensions.

TABLE 5.3 Typical coil widths for aluminum.

Typical coil dimensions		
Coil width (metric) (mm)	Coil width (US) (in.)	Comments
1000	39	
1250	49	
1500	59	
1524	60	
1829	72	Limited source
2438	96	Limited source

Aluminum is often provided in coils that have received a paint coating. The paint coating is applied at a secondary finishing operation, not the mill source. Aluminum coils can also be furnished with an embossed texture. This is also not a mill applied finish but a secondary finishing operation.

The coil can also be "decoiled" to specific, predetermined sheet lengths. As the coil is unwound, the sheet is passed through a leveling station to remove some of the internal stresses imparted by the coil. A shear, programmed to cut the aluminum coil at a predetermined length, evenly slices through the ribbon of metal and deposits the sheet onto a stacking device. A skid of sheets can be established by maximum weight, number of sheets per skid, or other preestablished criteria. All the sheets are decoiled and stacked onto the skid in the same direction with the finish side up.

Often the aluminum sheets are coated in a thin protective plastic film as they are decoiled. Special papers can also be interleaved between the stacked sheets. These protect the surface of the sheets from rubbing against one another during transport. The surface of the aluminum sheets when they arrive at the factory for further processing should be pristine, dry, and free of scratches or mars.

Widths of sheets are subject to the coil width. Varying the width from the coil width, requires special blanking to specific sizes to be performed at a secondary processing facility. At this facility the large coil may be decoiled to smaller coils, slit into narrow coils or blanked into sheets of a specific dimension. Shipping of lengths longer than 4 m will require special crating and handling to avoid damaging the material. On long lengths it may be advisable to ship as coils, then process the coils locally to avoid damage from shipping. Many facilities that work aluminum in long sheets have arrangements to decoil and level the sheets. Roll forming operations often have systems in place to decoil in line. Most of these facilities have lower limits on the size and weight of the coil and require the coils to be decoiled into smaller coils for handling.

Decoiling operations often include in-line flattening or straightening of the coil. This step is critical to ensure consistency across the sheet and to adjust the internal stresses in the thinner sheet material (Figure 5.8).

FIGURE 5.8 Basic leveling line taking metal from a coil and feeding a roll former.

FOIL

Aluminum is one of the few metals that is readily available in a multitude of foil thicknesses. The thicknesses available are in increments of tenths of millimeters or decimal points of inches. Foil is considered to be aluminum that is rectangular in cross-section provided as a coil. Aluminum foils are available in 0.005 mm (0.002 in.) to 0.24 mm (0.0094 in.) thicknesses, which, in the United States, laps over into thin sheet thicknesses. Household aluminum foil, for instance, is approximately 0.016 mm, but it can be supplied in heavy-duty form as thick as 0.24 mm (0.0094 in.).

This ability to be rolled very thin and still maintain some degree of tensile strength, allows aluminum foils to be used alone or in conjunction with insulation material. Aluminum foil is typically produced from continuous casting of aluminum and rolling the cast billet to a very thin ribbon.

Foils can be ordered in various tempers from fully annealed to H18 full hard tempers. The alloys available are the commercial pure A91xxx series, as well as A93xxx, A95xxx, and A96xxx. There are other specialized alloys that are available as well. Aluminum's ability to reflect long wavelengths, such as infrared radiation, allow thin foils to be used to keep heat in, such as in the case of food wrappings, but also to reflect it out as an insulating material.

Foils are usually bright in appearance due to the way they are created in polished pressure rolls. The rolls need to be highly polished as they compress the ribbon of aluminum without galling or tearing the surface. Often the foil is rolled thin in two back-to-back ribbons. This reduces the foil ripping in the process. The back-to-back surfaces has a mill, as-fabricated finish, a nonuniform surface that can vary from coil to coil or within a coil of the thin foil material.

Aluminum foil is available in the following finishes:

- Both sides specular and bright
- One side specular and bright with other side mill finish as-fabricated

There are various types of aluminum foils developed for particular industrial uses such as printing and embossing for packaging or for laminating to other materials. These foils can have different finishes and reflectivity (Figure 5.9).

Standard widths of foil are:

Foil width (Metric) (mm)	Foil width (US) (in.)
152	6
203	8
300	12
450	18
500	20
900	36
1220	48

There are numerous industries that use aluminum foil laminated to various substances. Thin aluminum foil can provide the thermal reflective properties in a highly malleable form to reflect radiation. From a fabrication standpoint, foils are not the easiest to work with because they easily tear. Folding and shaping is performed by hand or in shaping and crimping rolls. Their lack of rigidity makes it difficult to achieve and maintain tolerances.

Conventional joining methods, such as welding, riveting, and bolting will not work with foils. Joining can be done by rolling and crimping edges together to create low-stress joints. You can use adhesives to apply foils to a rigid substrate.

FIGURE 5.9 Example of aluminum foil surface finish.

CLAD ALUMINUM: ALCLAD

Clad aluminum is the coating of a one alloy of aluminum over another. The name for this is Alclad, a trademark owned by Alcoa. Usually it involves the protection of a high strength aluminum alloy with a higher purity, more corrosion resistant alloy. The higher purity alloy acts as a barrier for the more corrosion-prone high-strength alloy. Alclad involves casting an ingot of the high-strength core aluminum and placing it while hot in a mold lined with the more corrosion-resistant aluminum alloy. The metal is bonded by passing it through high-pressure rolls and essentially pressure welding the two together. Heat treatment processes required to bring the strength up on the core metal adds to the diffusion of the two aluminum alloys. The bonding is performed cold and is also referred to as roll bonding. The process was first used back in the 1920s as a surface material for a naval blimp called the ZMC-2. Since then it has been an important method of providing protection to high-strength aerospace aluminum alloys. Highly pure forms of aluminum are coated over the wrought alloy A92024 to provide corrosion protection while maintaining the strength characteristics of the A92024 aluminum.

The cladding aluminum alloy is typically more anodic than the core material and thus some galvanic protection occurs. The layer of more corrosion-resistant aluminum is only 2–5% of the

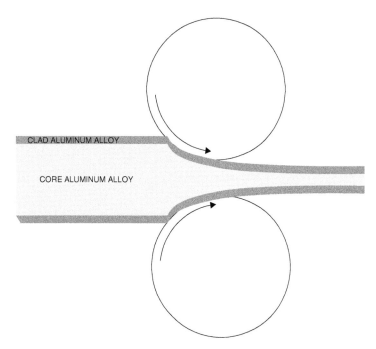

FIGURE 5.10 Roll bonding of aluminum onto an aluminum core.

overall thickness of the core metal and can be applied to all sides. Alclad alloys are very resistant to corrosion due to the high purity of the outer coating (Figure 5.10).

Alclad is performed on wrought forms of aluminum. Sheet, plate, wire, rod, and tubing can be roll-bonded with a more corrosion resistant outer layer over a stronger inner core. The outer coating is continuous, free of pores, and forms a strong barrier against corrosion attack of the core metal.

Alcad coated aluminum plate was used on the Museum of Science and Industry in Tampa, Florida, where it has provided superior corrosion resistance for decades of exposure. Even after 20 years of exposure the surface looks bright and reflective.

The finish of alclad aluminum is usually bright and reflective. A smooth, reflective tone is imparted from the pressure rolls to the surface. Note, however, the coating is usually soft and easily scratched. It can be anodized to harden the surface, particularly for thicker sheet and plate material.

EXTRUSION

Extruded aluminum offers one of the most versatile design tools for creating infinite shapes and surfaces. The designer can place metal where it is needed for strength, function, or aesthetics. In a sense, it is an additive process in design. Metal is not removed by milling but designed into a single

FIGURE 5.11   Examples of various aluminum extrusions.

die that can be used to shape an infinite quantity of linear forms. No other metal has the ability to take on so many incredible forms with such ease as aluminum extrusion offers (Figure 5.11 and Figure 5.12).

Aluminum has a low melting point, which makes it ideal for extruding. The heated billet becomes soft and pliable, allowing pressure to force it through a profile die. Once extruded, strength and ductility can be established in posttreatment processes. All alloys of aluminum can be extruded. In the context of architectural and art fabrications, the A96xxx is the most common alloy series used for extruding. Within this alloy series, the following alloys are the most common. See Table 5.4.

Extrusions are often called profiles, because of the cross-sectional appearance and the cut shape of the steel die they are extruded through. The Mill will provide special billets in cylindrical cross-sections to an extrusion facility. The Mill will cast the alloy to the specific alloying constituents and produce a large billet of aluminum. Like the large block used to make sheet and plate, the billets need to have the impurities that rose to the surface after cast removed. This is called peeling.

TABLE 5.4 Common A96xxx alloys used for extrusions.

Alloy	Extruding attributes	Aesthetic attributes
A96005	Good structural strength. Good extruding properties.	Anodizes well. Good strength.
A96060	Excellent extruding attributes. Medium strength.	Good clarity and consistency.
A96061	More difficult to extrude complex shapes.	Higher strength. Poor surface quality.
A96063	Very good extruding attributes. Most common.	Anodizes well.
A96463	Good surface attributes. Good extruding attributes.	Chemical brightening, good clarity.

The cylinder of solid aluminum is cut to length then passed through a set of tools that remove the outer layer containing the impurities.

FIGURE 5.12 Typical aluminum extrusion packaging.

Nearly all aluminum extrusions are created using the direct extrusion process. The direct extrusion process involves heating a billet of aluminum to soften it while a hydraulic press applies pressure to a ram. The heated aluminum becomes plastic and is pushed through an opening cut in a steel die. The cross-section of the aluminum takes the shape of the opening in the steel die and a long linear "stick" of aluminum is brought out.

The beauty of extrusion is versatility in design and consistency in shape. Designing extrusions allows just the right amount of aluminum where required to meet structural conditions as well as shaping to achieve aesthetic characteristics. Screw guides can be established in the aluminum shape to locate and position assemblies. Connection features can be designed and extruded to enable multiple extrusions to interface to produce one large surface. Gasket holding grooves can be cut into the shape to incorporate other materials into an assembly. The ability to accurately create various dimensional forms is the hallmark of the extrusion process (Figure 5.13).

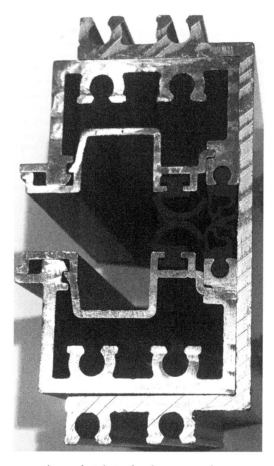

FIGURE 5.13 Example of incorporating gaskets into aluminum extrusions.

The process of extrusion is older than the metal aluminum itself. It was used on copper and lead prior to the commercialization of aluminum. However, today extrusions are nearly synonymous with the metal aluminum. Our modern glass curtainwalls are created with aluminum extrusions, and handrails, floor grates, expansion joints, and automotive trim are often produced from aluminum extrusion. Any shape that can be fabricated from a long linear straight form with a consistent cross-section can be economically and efficiently fabricated by extruding aluminum. Postshaping can be effective with aluminum extrusion but only after the initial extruding process and tempering process.

Extruded aluminum offers one of the most versatile design tools for creating infinite shapes and surfaces. The designer can place just the right amount of metal where it is needed for strength, function, or aesthetics.

EXTRUSION PROCESS

All extrusion processes begin with a cylindrical billet of solid aluminum. The billet is composed of a select alloy of aluminum created to specification at the mill source. The billet is often cut down from a larger cylindrical casting called a log. Billets are in lengths of 660 mm (26 in.) to as large as 1830 mm (72 in.) and can be in diameters of 76 mm (3 in.) to as large as 838 mm (33 in.) depending on the equipment set to receive them. The larger diameters require significant tonnage in the press used to push the heated aluminum through the die. The more common diameters are 178 mm (7 in.) and 254 mm (10 in.).

The billet is heated and the equipment, the press and die, are preheated. The billet temperature for the A96xxx alloys is brought up to around 496°C (925°F). These high strength alloys need to work at a higher temperature and this temperature needs to be maintained as the metal exits the die. Note the melting point of aluminum is 660°C so by heating the billet to 75% of the melting point, the metal is still in solid form.

Extreme pressure is applied to the end of the billet by means of a special hydraulic extrusion press. The pressure can range from 100 tons to as high as 15,000 tons depending on the equipment. This pressure pushes the hot billet up against the steel die. As the pressure builds, the semi-plastic aluminum flows through the die opening and conforms to the shape cut into the die.

As the hot extrusion exits the die it is run out onto a set of roller tables. A single billet may push out as much as 60 m at a time depending on the profile. As it is brought out on the table it is cooled by a water quench or air wash. The extrusion may twist due to differential cooling or out of balance stress due to the shape. It is cut to lengths to fit into a stretching device. The length is stretched 1–3% to eliminate the stress and straighten the extrusion. The ends may be trimmed to a finish standard length, typically 5–7 m. Following this it is placed into an oven to undergo artificial aging in the case of A96xxx. This achieves the required strength.

For example, you may see an extruded alloy specified as:

A96061—T651

This specification states the metal was solution heat treated, stress relieved by stretching, and artificially aged to maximum strength. Extrusions are long, linear lengths with variable shape enhancements. Stretching is the common means of adjusting the differential stress that may develop in the process of heating, extruding, and cooling. The result is a length of aluminum or a series of lengths all with precisely the same cross-section. The novel feature is the complexity of design this allows with a single die shape. The die can be utilized over and over to produce lengths of exactly the same profile shape. The key lies with the shape to be cut into the die.

DESIGNING THE SHAPE

The shape, also known as the profile, has several constraints that will affect whether the part can be successfully extruded, the rate of production and the ultimate cost.

Extrusions are defined into three basic categories, hollow, semi hollow, and solid. A hollow form has a perimeter enclosing a void or series of voids. Many extrusions for curtainwall systems are hollows, as are pipes and tubes.

A semi-hollow is not totally enclosed and has what are referred to as tongues, areas of perimeter metal that extend into the void but do not totally enclose the void.

A solid may have insets or protrusions from a central core. It could be rod-like in profile. But there is no significant void region. Structural forms such as I-beams and channels are solid extrusion forms (Figure 5.14).

These categories all are affected by the constraints and determine if features must be added or changed in order for a successful extruded part. The hollow forms require special dies and the flow of the metal must be understood as they pass through these dies to create the hollow form. All of the categories must fall within the process constraints.

CROSS-SECTION DESIGN CRITERIA

The first of these constraints is the overall cross-section, and whether the profile can fit within a given circle size. This constraint is referred to as the circumscribing circle size (CCS). The most common CCSs are 178 mm (7 in.) and 254 mm (10 in.). There are larger presses that can accommodate larger diameter billets and thus larger profiles, but these will require a significant increase in tonnage and will be more expensive. Additionally, larger size profiles require more care and work

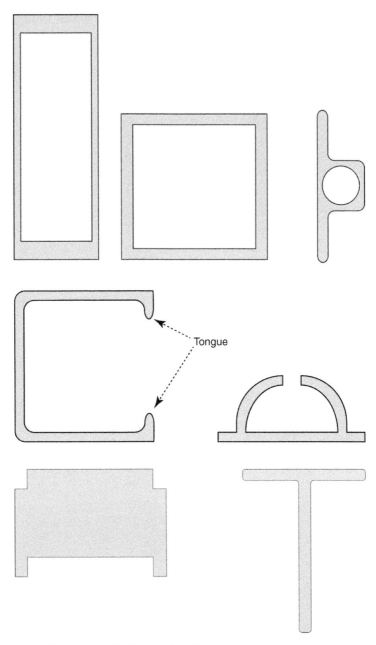

FIGURE 5.14 Examples of hollow, semihollow, and solid extrusions.

as they exit the die. They may twist more, may show more flaws induced from the die or variations in the billet. Therefore, if the design is going to incorporate a large single extruded form, it is highly recommended to work with the extrusion facility and get their input on the design (Figure 5.15).

If the design calls for a part that is larger than a given circle size or if the cost of using the larger die is extensive, consider two profiles and assemble the part from two separate extrusions pushed through a smaller circle size.

When the shape is created it is important to determine the greatest edge-to-edge distance of the shape, cross-sectional area of the shape and the perimeter measurement of the shape. This information is important in establishing the die that will be needed to create the shape and the adequacy of producing the cross-section.

Maintaining the mechanical properties is critical in many architectural extruded forms. The aluminum extrusion industry uses a simple rule called the extrusion ratio. The extrusion ratio is the cross-sectional area of the billet divided by the cross-sectional area of the profile.

Extrusion ratio = Billet cross-section area/Profile cross-section area

If the ratio is small, less than 10:1, then there will be some loss of mechanical properties as the relatively large shape is pushed through the die. You don't count the hollow part in the cross-section, only the profile and the area of metal that will be created.

You want a ratio between 10:1 and 35:1 for optimum results.

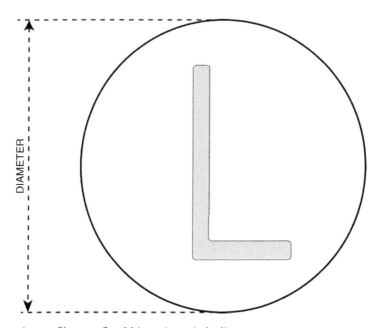

FIGURE 5.15 Extrusion profile must fit within a given circle diameter.

Billet cross-section		Area of billet		Range of optimal profile area	
76 mm	3 in.	182 cm^2	28 in.^{2}	5–18 cm^2	1–3 in.^{2}
178 mm	7 in.	993 cm^2	154 in.^{2}	28–99 cm^2	4–15 in.^{2}
254 mm	10 in.	2026 cm^2	314 in.^{2}	58–203 cm^2	9–31 in.^{2}
381 mm	15 in.	4558 cm^2	706 in.^{2}	130–456 cm^2	20–71 in.^{2}

For all extrusions, the extrusion producers examine the difficulty factor of producing a particular shape, regardless of the alloy. The difficulty factor is defined as the perimeter dimension divided by the weight per unit length.

$$\text{Difficulty factor} = \text{Perimeter of extrusion}/\text{Weight per unit length}$$

The higher the value the more difficult the shape will be to produce. It is not necessarily the die cost, this does have an effect, but here the complexity refers to difficulty of the aluminum pushing though a shape that has a lot of surface area. More tonnage may be needed and slower press time may be needed, resulting in a lower production rate. This would lead to higher unit cost.

EXTRUSION DIES

Most dies are made from hardened tool steel that has the cross-section profile cut into it. Tool steel used is a special steel alloy containing chromium and molybdenum. This steel, known as H 13 steel, has good strength when heated and can withstand thermal cycling from the extrusion process without cracking or developing excess wear. Cutting of the profile into the steel is performed by different methods but the more common one today is electrical discharge machining (EDM). Wear on the die, which will affect the finish surface of the extrusion, is controlled by nitriding the tool steel at the perpendicular surface where the aluminum passes through the profile die. Nitriding hardens the steel, and as aluminum passes this bearing surface, friction between the metals determines the flow rate of the extrusion. Extrusion companies renitride the bearing surfaces several times to extend the life of the die and keep the surface of the aluminum extrusion consistent (Figure 5.16).

In the process of extrusion there are multiple dies called support dies and are referred to as backers, bolsters, sub-bolsters, and feeders. For hollow dies a two-piece assembly is utilized in addition to the support dies, one to form the outside and one to form the inside. The backing dies and bolster dies are there to support the profile die while pressure is applied in the extrusion press.

These added dies have specific purposes. Feeder plates are positioned next to the billet. As a billet is pushed through the die stack and the extruded form exits, subsequent billets are placed into the press and welded at the feeder plate, allowing continuous extruding over multiple billets. This modern practice allows for a production process to manufacture parts and keeps the process competitive while reducing waste. Special feeder plates aid in achieving design tolerances by ensuring pressures and flow rate are consistent.

The feeder plate also is designed to move the heated and pliable aluminum to areas needed in the profile die. For example, if the profile extreme edge to extreme edge is greater than 90% of the

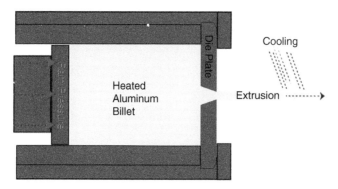

FIGURE 5.16 Basic extruding process.

billet diameter, the feeder is designed to equal out the flow of metal to the extreme edges. This helps produce an extrusion with good stress distribution as the flow rate through the opening is at an even rate.

When creating a die for a particular shape, there are several basic rules to follow. Essentially, the aluminum flows through the die in straight lines at a low velocity so the condition is considered laminar and is subject to physical constraints of fluid dynamics. The die walls that come into contact with the aluminum create friction. For large solid forms, the center flows at a different rate than the walls because there is no friction at the center. On intricate shapes the more friction on the sides slows down the flow and can create differential stresses as the shape exits the extrusion. Weight of the aluminum also plays a role. The heavier the cross-section unit is, the more difficult it is to extrude.

Rules of Designing with Aluminum Extrusions
- Minimize asymmetric shapes
- Balance wall thicknesses across the cross-section
- Minimize hollows
- Round edges
- Add grooves, offsets, ribs to aid in flow balancing
- Keep the perimeter/cross-sectional area ratio low
- High tongue ratio: width of opening / height of protruding surface
- Economy is achieved with quantity
- Review design with extrusion company
- Consider the right alloy and temper

Aluminum extrusions are a unique design form. A designer can incorporate strength while eliminating redundancy. At the same time features can be incorporated to accept fasteners, conceal seams, and develop specific aesthetic traits. In designing with extrusion, understanding the process and the constraints is important for success.

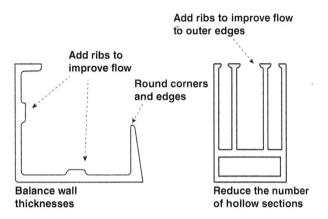

FIGURE 5.17   Improvements to extrusion design to aid in metal flow.

Extrusion profile design is intuitive when you consider how a viscous fluid wants to move through an opening. When the opening is out of balance the thicker region puts up less resistance than the thinner region and this causes imbalance in pressures. The larger area has less resistance, and this is where the metal will want to go.

Similar are sharp edges and corners, which put up resistance to the flow and should be avoided.

You can aid in the flow of metal through the die by adding grooves, webs, ribs, and rounding edges. This is not putting unnecessary metal into a design but allowing a design to be thinner and more detailed (Figure 5.17).

Creating functionality with extrusions can save time and cost. Adding indexing features, screw guides, and screw receivers allow for the joining of multiple extrusions into larger assemblies. Extrusions can be designed within tolerances that enable one piece to snap into another to produce a hollow shape without the higher expense of a hollow shape die. The possibilities are endless. Extrusions are one of the most compelling reasons to design with aluminum.

ALUMINUM PIPE AND TUBE

Aluminum is available in both pipe and tubular form. It is important to understand the difference between pipe and tube. Pipe is used to transport fluids within them while tubing is structural. This means that dimensions such as wall thickness and outside diameter are critical for tubes, while the inside dimension is critical for pipes. The tolerances for the manufacture of tubing are much stricter than for pipes.

Pipes are categorized as vessels that transport fluids and the outside dimension is considered "nominal," which can be different from the measured outside dimension. Generally, pipes are listed as schedule 10, schedule 40 or schedule 80, which defines the approximate inside diameter. The outside dimension for different schedule pipes is roughly the same, the inside dimension changes

TABLE 5.5 Differences in actual dimensions of schedule pipe.

Pipe schedule	Outside diameter				Wall thickness		Inside diameter	
	Nominal		Actual					
	mm	in.	mm	in.	mm	in.	mm	in.
10	25.4	1.00	33.53	1.32	2.79	0.11	27.94	1.10
40	25.4	1.00	33.53	1.32	3.30	0.13	26.67	1.05
80	25.4	1.00	33.53	1.32	4.57	0.18	24.38	0.96
10	50.8	2.00	60.45	2.38	2.79	0.11	54.86	2.16
40	50.8	2.00	60.45	2.38	3.81	0.15	52.58	2.07
80	50.8	2.00	60.45	2.38	5.49	0.22	49.28	1.94
10	76.2	3.00	88.90	3.50	3.05	0.12	82.80	3.26
40	76.2	3.00	88.90	3.50	5.49	0.22	77.98	3.07
80	76.2	3.00	88.90	3.50	7.62	0.30	73.66	2.90

as the wall thickness defined by the schedule changes. As an example, for three sizes of pipe, you can see in the following Table 5.5 that as the schedule changes, so does the inside diameter and the wall thickness.

Pipe often has a rougher surface finish because it is not meant to be for aesthetic purposes and is stored, handled, and transported in ways that will mar the surface. Most aluminum pipe is provided in A96061—T6. But pipe is also provided in A96063 and A96101. Pipe is considered as having a circular cross-section and is always listed as nominal outside diameter and schedule.

Aluminum tubing comes in round and rectangular cross-sections. Unlike pipe, the dimensions for wall thickness and outside diameter are accurately stated for determining the structural properties of the shape. The finish on tubing is superior to that of pipe; however, when ordering tube, it is important to relay the quality expectations to the supplier to avoid scratches and mars. Tubing is used for railings, art features, and other architectural features that need engineered performance criteria. Tubing can have a small cross-section or very large cross-sectional shapes. The limitation is in the manufacturing process. Most tubing is produced by extrusion press.

There are several common alloys used for tubing: Alloys A96060, A96061, A96063, and for more strength, alloy A92024. Tubing comes in circular, square, and rectangular cross-section. Circular and square cross-sections have wall thicknesses that are consistent, while rectangular tubes can be consistent or can vary, such as where the long side is thinner than the short sides. Both square and rectangular tubing is available in rounded or square corners.

When anodizing tubing, there can be color differences due to metallurgic changes as the extruded edges come together to form the tube. Some extrusions bring the edges together at a corner, where they are fused together during the extruding process to achieve the shape. Others may come together in the middle of a wall. When anodizing, this seam can show up as a line of a slightly different color.

ALUMINUM ROD AND BAR

Aluminum is stocked in rod and bar shapes. Cross-sections are solid forms of circles, squares, rectangles, and hexagons. Various sizes are available. Usually the finish is not special and can be marred and possess minor scratches. Streaks and lines from the dies used to create the stock forms may also be apparent. Rod and bar stock are used for various industrial processes. Further machining is the normal path for this material form. Posttreatment options may be needed to arrive at an appropriate finish. Typical alloys are A96061 for the bar and rod shapes and A92024 for the hexagonal shape (Figure 5.18).

ALUMINUM STRUCTURAL SHAPES: ANGLES, CHANNELS, TEES, AND I-BEAM SHAPES

Aluminum is also available in many structural forms such as angles, channels, tee shapes, and I-beam shapes. The alloys used are typically A96061 and A96063. They are available in various tempers to achieve strength requirements. Alloy A96063 proves a better surface finish.

These shapes can have either square edge for aesthetic purposes or rounded for structural performance. The I-beam and channel shapes can be up to 305 mm, (12 in.) in depth and the angle shapes can have legs as large as 203 mm, (8 in.).

FIGURE 5.18   The *1-3-5-7-11* sculpture, made from solid aluminum bar, by Sol LeWitt.

The structural shapes are extruded and then tempered to achieve strength requirements.

The mechanical properties of some common aluminum structural shapes can be obtained from published data or calculated by the cross-sectional shape.

The limitation with aluminum as a structural component is deflection. The modulus of elasticity is one third that of steel. Yield and tensile strengths can be increased by heat treatment and alloy selection, but it is the elongation that remains a challenge when designing with aluminum (Figure 5.19).

When using aluminum, consider improving the moment of inertia to overcome the reduced elasticity. To do this, consider planning for deeper sections or place more mass at the outer edges of the shape (Figure 5.20).

Aluminum can be subject to fatigue. Fatigue occurs when the aluminum form undergoes cyclical loading. When the load goes beyond the plastic limit and the metal deforms it will undergo work hardening. When the load is removed the metal will contract in length slightly. When the load is

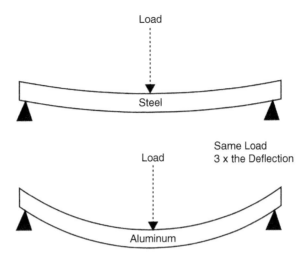

FIGURE 5.19 Deflection of aluminum in relation to steel.

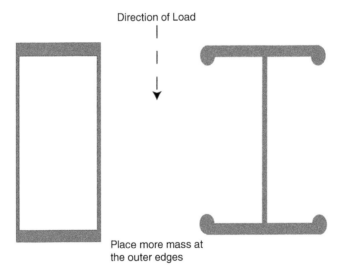

FIGURE 5.20 Increase the moment of inertia to offset deflection in aluminum.

reapplied, slight expansion will occur. As this is repeated in the cycling of loads, the work hardening accumulates to the point where it no longer can undergo plastic deformation and a crack will develop. As loading continues, the crack grows (Figure 5.21).

When designing to resist fatigue use a smooth, polished aluminum surface or shot peened surface. When the mill finish is used, it has small microscopic impressions that can lead to crack development. For aluminum, the quality of the surface is as important as the alloy choice.

FIGURE 5.21 Fabrication facility. Structural fins made from A95005—H32. Designed by Crawford Architects.

Anodizing the surface will reduce fatigue resistance. The anodized coating is very inflexible and will crack under loading. These cracks can grow as loading is applied.

Shot peening the surface of aluminum with stainless steel shot will harden the surface and improve fatigue resistance by putting compressive stresses into the metal. These will work to counter the stresses from fatigue loading.

ALUMINUM WIRE

Aluminum wire is available in different sizes. The majority of aluminum wire goes into electrical usage but is often used for tie wire due to the ductility of the metal. Wire is sold in spools or coils by weight or approximate length.

Aluminum wire comes in various alloys but the most common are the pure forms due to the ductility. Alloy A95052 is another used to manufacture wire, in particular wire destined for the manufacture of wire mesh or wire cloth. It has better strength than the A91xxxx alloys. Thicknesses start at 0.025 mm (0.001 in.) diameter up to 2.29 mm (0.090 in.) diameter.

ALUMINUM WIRE MESH

Aluminum is an excellent material for wire mesh. Wire mesh, also called wire screen and wire cloth, was once restricted to forms such as sieves, insect screening, and fencing for small animals. More recently, wire mesh has found a place in architecture and art, due to its light weight and corrosion resistance.

Woven wire mesh is expressed in wire size and mesh count. Mesh count represents the number of openings in a square cm or square inch, depending on the manufacture and country of origin. In the United States, the alloy typically used is A95052 and the expression of mesh size is given as openings in a square inch. For example, a 2 × mesh size would represent 0.5 in. squares or 4 per square inch. They can go as small as 200 × 200. This is a very tight screen of 40,000 squares per square inch. There is also what is referred to as "off count mesh" which calls for different number of openings when measured in one direction versus the other.

The wire size is given in gauge of wire, which corresponds to a specific diameter. In the United States, the Brown and Sharpe wire gauge (a.k.a. AWG, American Wire Gauge) is a logarithmic value, where increasing number of values correspond to smaller and smaller wire diameters.

For wire meshes used in architecture, there are a number of custom forms in use and more are being developed by various manufactures. Plain weave, interlocking, interlocking and crimped, and twilled weave are just a few of the woven wire mesh patterns available.

ALUMINUM EXPANDED METAL

Expanded aluminum is a screen-like material made by piercing a sheet of aluminum and making a small slit. The pierced sheet is then stretched to create a pattern of small openings, usually in the shape of a diamond (Figure 5.22).

No metal is removed in the process and the sheet becomes wider or longer as the diamond shaped opening is expanded. Typically, the openings have a slightly raised "lip" created as the stretching occurs. This form is known as standard expanded metal or formed expanded metal. This formed sheet can be further processed to flatten out the lips, referred to as flattened expanded metal.

The use of expanded metal in art and architecture has had wide acceptance in the last few decades as it has moved from an industrial, utilitarian material to an aesthetic design material. Aluminum is well suited for expanded metal sheets due to the ductility of the metal and excellent corrosion resistance. The edge of the pierced metal is exposed to the environment, but aluminum will not present an edge that is susceptible to deterioration in the same way other coated metals will.

Expanded aluminum sheet or plate can be ordered with a diamond pattern running lengthwise or across the sheet. The diamond pattern can be staggered or straight line. The diamond can be created in a multitude of sizes from very small on thin sheet to very large.

FIGURE 5.22 Expanded and bumped aluminum gates, designed by Herzog and deMueron.

Other decorative patterns are also available from different manufacturers of expanded metal. Availability, sizes, and thicknesses of more custom openings should be investigated directly with the manufacturer (Figure 5.23).

When working with aluminum expanded metal, it is important to recognize the way it interplays with light. In particular, the finer, smaller screen-like material. In bright lighting conditions, the, material appears solid, like an opaque textured surface. Adding shape to the surface can create high and low spots resembling fabric or stone. Looking from a darker region though the material, it takes on a totally different appearance, like a diaphanous shroud. The condition changes back and forth as the lighting conditions change from one side to the next (Figure 5.24).

Thin aluminum expanded metal meshes can be more brittle than thicker, larger aluminum expanded metal. Forming an edge on the metal can be difficult. The standard expanded metal is not flat and in a conventional press brake keeping the edge can throw off the accuracy and make

FIGURE 5.23 *Winds of Aphrodite*. Layered and bent expanded metal. Designed by Crawford Architects and artist Suikang Zhao.

FIGURE 5.24 Looking from inside to outside on the expanded mesh.

achieving tolerances difficult. When the aluminum is pierced and stretched there is localized cold working than can make the aluminum less ductile. If the fold is a sharp bend, a crack can occur and propagate along the edge. On the flattened expanded metal, forming does not pose a problem.

PERFORATED ALUMINUM AND EMBOSSED ALUMINUM

Aluminum sheet is an ideal material for perforating and bumping. The metal is softer, and the lower yield requires less energy. This leads to less localized stresses being imparted to the surrounding metal. Because aluminum has such good corrosion resistance, the edges of the holes do not corrode. Painted aluminum can be perforated with less concern about premature failure of the paint coating. Purchasing prepunched aluminum sheets in different gages and alloys is similar to buying expanded aluminum sheet. There are many toll perforation companies that can provide different patterns of holes and different thickness of aluminum. As with expanded metal, aluminum offers an edge that, where pierced, is the same basic makeup as the balance of the material. The thin protective oxide layer will grow to offer corrosion protection to the edges (Figure 5.25).

Aluminum can be custom perforated. Custom perforation involves piercing with a tool and die controlled by computer-driven equipment. Turret punch and similar equipment takes less pressure to pierce an equivalent hole in aluminum than in steels. Additionally, there is less work hardening, so the metal remains flat as the piecing occurs.

Along with this, less pressure is needed for embossing or bumping of the surface of aluminum. The plastic behavior of the metal allows aluminum to be formed deeper on embossing rolls or with computer numerical control (CNC) equipment (Figure 5.26).

FIGURE 5.25 Perforated aluminum plate, designed by WRNS Architects.

224 Chapter 5 Designing with the Available Forms of Aluminum

FIGURE 5.26 Aluminum with dual pattern of perforation and bumping, designed by SERA.

CAST FORMS

Cast forms of the metal aluminum are common today in all industries. The first uses of this newly discovered metal in the late 1800s and early 1900s was for cast ornamentation. In the early years, casting ornamental features and shapes was the main form. The top of the Washington Monument was cast, as was the spire adorning the Smithfield Church in Pittsburgh, Pennsylvania.

Aluminum can be cast using several common industrial techniques:

Sand casting
Permanent mold casting
Die casting

Investment

Centrifugal

Aluminum used in architectural is predominantly cast in sand casting, permanent mold casting, or die casting. An artist may, on occasion, use the method of investment casting, also known as lost wax technique. Centrifugal casting is used to create smaller parts for industrial uses. Centrifugal casting is good for creating repeated parts that have symmetry around an axis. The spinning die receives the metal by means of centrifugal force (Table 5.6).

Aluminum offers several advantages to other metals used in the foundry for casting. Aluminum melts at a considerably lower temperature range than other metals, so energy costs are lower. Aluminums lower density makes it easier to handle. The light weight allows for more flexibility in how the metal is handled in the foundry when pouring smaller parts. The lower density produces lower mold pressures; this allows for lighter weight equipment. Aluminum, however, is more prone to taking on impurities in the casting process. This can lead to porosity in the casting.

There are several factors to consider when casting aluminum and these should be considered both in the design and in the selection of alloy.

1. *Fluidity*. This is the ease in which a metal flows. Good fluidity allows for fine detail and thinner sections and reduction of rejected castings due to incomplete mold filling. The foundry considers how far the metal must flow in the mold before solidification begins. Some alloys have better fluidity than others and the foundry will help determine this. There are a number of critical variables that play a role in the fluidity of an alloy in a particular mold. Higher silicon content tends to improve fluidity up to a point. Other alloying elements tend to reduce the fluidity.

2. *Shrinkage*. Shrinkage is a property of aluminum casting and must be considered in design of the mold and alloy choice. Shrinkage is a term given to the change in dimension as the liquid aluminum undergoes solidification in the mold. Different elements in the alloys solidify at varying rates and this will have a great effect on the soundness of the casting. As aluminum castings begin to cool, they solidify at a linear rate. Castings are more linear, and the mold designer can somewhat predict the shrinkage and solidification. On complex shapes, predicting how the part will shrink is difficult. Always consider test runs to enable adjusting the mold to accommodate changes that may be encountered. Mold adjustments include adding risers and vents, tapering sections, and even adding chill bars to control where the solidification needs to occur first.

    Shrinkage is designated into three range classifications (narrow, medium, and wide), each requiring special mold design. A narrow shrinkage range solidifies from the walls of the mold inward and from areas of small mass to large mass. Narrow shrinkage range alloys require large risers. This keeps the thermal mass larger and, in a position, to feed into the casting. The medium range is the most forgiving of the three classifications. Smaller risers are needed for medium range alloys. The wide range of shrinkage makes it difficult to achieve a good

TABLE 5.6 Comparisons of cast methods used on aluminum.

Casting method	Relative cost	Process speed	Size	Surface quality	Dimensional accuracy	Common alloys
Sand	Low	Slow	Small to large	Poor to medium	Low	A03190 A33550 A13560 A03570 A04430 A05200 A07130
Permanent mold	Medium	Fast	Small to large	Medium to good	Good appearance	A03550 A03560 A13560 A13570 A03660 A24430 A05130
Die cast	High	Fast	Small to medium	Good	Very accurate	A03600 A03800 A13800 A04130 A05180
Investment	Medium	Slow	Medium	Good	Low	A02080 A03080 A03190 A03550 A03560 A04430 A05140 A05350 A07120
Centrifugal	High	Fast	Small to medium	Good	Very accurate	A13560 A17710

sound casting. For wide ranges, you need to design the part, so the thermal masses are as uniform as practical. For wide shrinkage range alloys, the edges will solidify at the same time as the rest of the casting.

3. *Dross formation.* Dross is the term given to the oxidized aluminum and other elements that float to the top and appear as a fine flaky film. It will solidify on a casting. It can be removed by blasting the finish casting but often pits or voids are left behind. In designing the mold for casting, keep the most visible aesthetic surface at the bottom of the mold.

4. *Temperature of pour.* A good design avoids hot spots in the casting that will have an effect on the metallurgy and physical properties. Sharp, inside corners on the mold will cause the metal to stay hotter for longer.

Each casting method has its unique characteristics and demands depending on the design requirements, metallurgical characteristics, and production demands. Economy plays a large part in which casting method to use. Rates of solidification of the metal is a major metallurgic consideration with each casting method. Sand casting has a slow freezing rate, while in permanent mold casting and die casting the metal freezes more rapidly. This is due in part to the mold itself. Sand casting can insulate, keeping the molten metal warmer, longer.

SAND CASTING

Sand casting of aluminum has low tooling cost, making it the most economical method of aluminum casting. Sand casting is straightforward, make a mold out of green sand and pour the molten metal into the cavity. Gravity pulls the molten metal into the mold and it flows to the edges. Casting with aluminum in sand is similar to casting of other metals into sand. Achieving a good surface finish, however, may be more difficult than other, costlier, methods, due to the surface finish of the metal created by the sand mold.

Sand casting offers the ability to cast large sections as compared to other methods. This casting method is used in low volume production processes. Sand casting of aluminum most often occurs in green sand. Cores are also of green sand. These are premade and baked to remove moisture then set into the mold. In certain instances, the entire mold may be baked to produce a "dry mold" free of moisture. Moisture can lead to unwanted porosity. The aluminum is gravity fed into the mold and the sequence of feeding the metal is designed by the foundry to establish a progressive solidification of the casting (Figure 5.27).

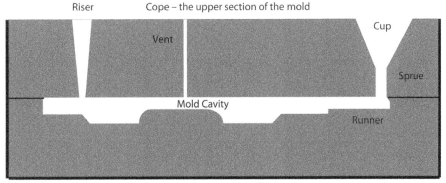

FIGURE 5.27   Cross-section of a typical sand cast mold.

PERMANENT MOLD CASTING

The molds used in this casting method are made of metals such as cast iron or steel. They are gravity fed molds that perform in a similar way to the sand cast process. Dry sand or steel casting can make up the cores used in these molds, particularly when the finish is not critical. Permanent mold casting produces better surface finishes than sand molds which tend to be grainy and lack smoothness. Dimensional tolerances are tighter with permanent mold casting than with sand casting methods. Permanent mold falls between die casting and sand mold casting in mold cost and finishing production costs.

DIE CASTING

Die casting is a process well suited for low temperature metals. The metal is forced into the mold under pressure. The molds are often water-cooled metal dies. Die cast aluminum is most commonly provided in the as cast temper, F. Die cast shapes do not lend themselves to strain hardening.

Die casting involves greater cost in the mold design and fabrication but the surface finish achieved is superior. This casting method is best suited for multiple parts. Die casting requires placement of the aluminum into the die while under pressure. The surface finish of the aluminum matches that of the steel die the aluminum is pressed into. Some post trimming may be required but cleanup is minimal.

Double die casting is another common method used to achieve a quality cast part. The part is first cast, then stamped to achieve tight dimensional control and surface quality (Figure 5.28).

Casting is influenced by several factors that will determine the best method to use. Starting with the design of the part to be cast will help determine the most appropriate process, the alloy to use, and the expectations for the final finish.

Factors	Sand casting	Permanent mold casting	Die casting
Cost of process	Low cost	Moderate cost	Highest cost
Production rate	Low	Fast	Fast
Size of casting	Small to large	Limited size	Limited size
Shape	Complex	Less complex	Less complex
Core	Sand	Steel	Steel, simple forms
Dimensional control	Rough	Good	Good
Surface	Rough	Good	Very good
Porosity	Low	Low	Medium
Cooling rate	Slow	Slow to moderate	Rapid
Grain	Coarse	Fine	Very fine
Strength	Low	High	Very high
Versatility	Very good	Very good	Excellent
Repeatability	Poor	Good	Very good
Post cleanup	Moderate	Minimal	Minimal

FIGURE 5.28 Double die cast aluminum ceiling, designed by Tange Associates.

ALUMINUM FOAM

Aluminum foam is a relatively new product development and has found extensive use in creating lightweight, high impact backing in the transportation industry. It has found uses as fire walls, acoustic dampening, and high impact stress distribution systems. It is fully recyclable.

FIGURE 5.29   Aluminum foam walls in the 9/11 Museum, designed by Snohetta. Image of closed- and open-cell foam.

In art and architecture, it is finding use as an intriguing surfacing material. The interior surfaces of the 9/11 Museum in New York, designed by Snohetta, are clad in aluminum foam material created into panels.

Aluminum foam, also called SAF for stabilized aluminum foam, because ceramics are mixed into the molten aluminum as gases are introduced. This stabilizes the internal bubbles and the aluminum solidifies as the other layer of an interconnected series of bubbles (Figure 5.29).

Aluminum foam is available in different densities and thicknesses. The strength to weight ratio is increased dramatically due to the way the cellular structure adds strength to the form. There is less than 25% by volume of base aluminum material; the rest is air. The surface can be adhered to flat sheet material to provide stiffness. It can be flattened to create a smoother surface. The cells are closed so the material will float on water. It can also be created with an open cell structure. It is available in the as-produced mill finish.

CHAPTER 6

Fabrication

Start by doing what's necessary; then do what's possible; and suddenly you are doing the impossible.

—St. Francis of Assisi

Aluminum is the metal designed for the fabricator. Its malleability is unmatched and it can be hammered, stretched, sliced, and bent to create the designs of modern civilization. The major attributes of strength-to-weight ratio, stable and predictable surface, and ease of manufacture make aluminum the material of choice in many industries. As a cladding and structure for transportation vehicles, as easy-to-maintain forms for household goods, even as computer housings, and as frames and skins for our buildings, aluminum is one of the most versatile materials known to mankind.

The ductility of aluminum affords the fabricator the ability to stamp, shape, and form sheet and plate with less energy than most other metals. Fabricating with aluminum requires skill and finesse. The softer alloys—A91xxx and A93xxx alloys—can have tempers that enable hammering and shaping of aluminum in ways similar to copper. Other alloys offer good strength without excessive springback, a condition experienced with forming higher tensile strength metals, such as stainless steel.

Aluminum can be stretched thin by hammering and deep drawing. It will work harden and can tear if the shaping is severe or repeated, but with some amount of interstitial annealing the strain buildup is relieved and further shaping can occur. Aluminum will work harden as it is cold worked, similar to the methods used for other metals, but aluminum reaches the point where it can become brittle and crack as more strain is induced.

Annealing aluminum involves heating it to near its melting point. Allow it to air cool. Rapid cooling by quenching in water can make the metal brittle. You want to ease it back, unlike other metals. The ductility of aluminum is one of the greatest assets to the fabricator of formed parts.

Understanding the importance of tempering and cold working is necessary when fabricating intricate shapes out of aluminum.

The modulus of elasticity, the measure of stiffness of a material, is approximately 70 GPa (10,000 ksi), depending on the alloy. This is significantly different from other architectural metals, such as stainless steel. For a given alloy the modulus of elasticity is the same across the tempers induced on the alloy. Increasing temperature will decrease the elasticity of aluminum. Elasticity of a material is an important engineering property used to determine how a material will act under loading. See Table 6.1.

The modulus of elasticity, also called the Young's modulus, relates to how an object will change in length when subjected to a load. When the object is under tension, it will elongate in length. When subjected to compressive forces, it can shorten. Stiffness is resistance to change in shape while elasticity is the ability to return to the original size and shape. The modulus of elasticity indicates a material's ability to deform and recover as opposing loads are applied to a given cross-section of a shape. Essentially, it is the slope of the straight-line portion of the stress-strain curve of a material. The steeper the slope, the stiffer the material; the stiffer the material, the more it resists change in shape, and the greater the modulus of elasticity.

The modulus of elasticity is an important mechanical property of solid materials (Table 6.2). When a material deforms under load it is actually absorbing energy and stores it in the form of elasticity. If the load permanently deforms the material, this energy is dissipated plastically. The strain for a given stress applied to the material is not dependent on the size or shape of the material; it is always the same. The deformation that occurs is dependent on the geometry of form.

Aluminum has a modulus of elasticity that is lower than that of steel and even lower than that of copper and copper alloys. The advantage comes with the weight of the metal versus its stiffness and strength. The modulus of elasticity is an important constraint to designing with aluminum. Aluminum will deflect elastically more than most metals and has lower maximum strength. Aluminum has good ductility and toughness, the ability to absorb energy before it fractures. Aluminum's stress-strain curve is flatter than for other materials. It lacks stiffness and strength, but this gives it toughness and the ability to be shaped significantly before cracking. This makes aluminum

TABLE 6.1 Modulus of elasticity of various aluminum alloy families.

Alloy	Modulus of elasticity	
A92xxx	73 GPa	10,588 ksi
A93xxx	69 GPa	10,000 ksi
A95xxx	70 GPa	10,153 ksi
A96xxx	69 GPa	10,000 ksi
A97xxx	72 GPa	10,443 ksi

TABLE 6.2 Modulus of elasticity of various materials.

Material	Modulus of elasticity (GPa)
Acrylic plastic	3.2
Douglas fir	13
Aluminum	70
Copper	115
Cartridge brass (C26000)	110
Monel 400	180
Lead	44
Tin	30
Zinc	99
Steel	200
Stainless steel	193
Titanium	103
Glass	50–90
Polycarbonate	2.6

very effective for resisting shock loads: the sudden loading of an object and how well a material absorbs elastic energy before it cracks. Couple this with the lightweight aspect of the metal, and a design can be made thicker with the addition of aluminum, to accommodate loading without adding significant weight.

Aluminum is the metal designed for the fabricator. It's malleability is unmatched. Aluminum is one of the most versatile materials known to mankind. The ductility of aluminum is one of the greatest assets.

CHALLENGES WITH ALUMINUM FABRICATION

There are several challenges presented with aluminum fabrication. Aluminum failure during fabrication is largely caused by necking or loss of ductility. The metalwork hardens and can fracture as the metal is stretched beyond certain limits. Ductility fractures occur in aluminum alloys where discontinuities in the aluminum become aligned and are unable to overcome the stress as the stain accumulates around inconsistencies.

Microcracking at Bends

For most forming operations of aluminum, small microcracks may be apparent on the surface where the less flexible oxide separates as it is stretched. Thin aluminum sheet metal can be readily formed using a press brake or roll formed in a series of progressive dies, but as the thickness approaches the extremes of sheet material, 3 mm (0.125 in.), tight forming can produce tiny cracks along the bend line. Anodized aluminum will fracture along the edge when formed. The thickened aluminum oxide that develops on the surface from anodizing is not as flexible as the base metal and when shaped, punched, or drilled, small cracks will form where the edge shears (Figure 6.1).

This is no cause for alarm and on most instances these cracks are not readily visible. The underlying aluminum in the crack develops the protective oxide and nothing becomes of it. On dark surfaces, such as prepainted or pre-anodized sheet, the cracks can be visible, and these may be aesthetically objectional. For painted surfaces, compatible touch-up paint can be wiped over the

FIGURE 6.1 Microcracks at bends.

cracked area then wiped off before it dries. The touch-up paint goes into the crack and the contrasting color is no longer visible. On anodized surfaces this "repair" using paint can also be performed. Essentially you are removing the unappealing contrast by covering the lighter color with paint. It is important to remove all excess paint outside the crack.

Galling

Aluminum has a tendency to gall. Galling is a term used to describe the adhesion between two surfaces as they pass over one another. Each surface may appear smooth, but microscopically there are high and low points known as surface asperities. These microscopic anomalies on the two surfaces can adhere to one another as they attempt to slide over each other. Microscopic high points crash into each other and these disparities plastically deform and engage in the other surface. The surface's asperities dig into one another and can rip up small strands or clumps from the opposing surface.

Galling is dependent on a material's surface hardness and smoothness. Anodized aluminum surfaces are significantly harder than unanodized surfaces and the oxide that develops is smoother and more even than an unanodized surface. Galling is reduced significantly on anodized surfaces. Anodic coatings are applied to aluminum surfaces expected to slide over one another.

Surface Marring

The relative softness of aluminum exposes it to easy scratching and surface marring.

Marring occurs when an edge runs over the surface of a sheet and leaves dark, contrasting streaks. Marring is related to galling in that the condition is a surface condition caused by aluminum rubbing over aluminum. Temporary protective coatings on the metal are advised. Thin plastic films or paper interleaf is a common added protective measure to protect the surface. Avoid dragging aluminum over other aluminum surfaces, otherwise a dark mar line or damage induced from galling can occur.

HANDLING AND STORAGE

In the fabrication process of aluminum there are several important steps to be closely followed to achieve success and reduce rework or replacement. The first critical step is handling and storage of the metal. Natural, unanodized aluminum is soft and can easily be scratched. No matter which form—extruded, bar, tubing, flat sheet or plate—aluminum surfaces will scratch more readily than other metals. Being passed across worktables, banded with steel bands, or dragged over other substances that have a hardness superior to aluminum can all cause scratches.

There is one side prime on sheet and plate material unless you direct the mill or supply house otherwise. "Both sides prime" is a special order condition, although aluminum sheet and plate often have superior reverse side quality, due to the way the mill manufactures the metal and skids the metal. The point, however, is that all sheet and plate metals have a prime side. This means the top surface is maintained and considered the prime side and should always be free of scratches of any kind. The metal can be ordered with a thin plastic sheet on the face side to further protect the prime side of the metal during fabrication.

Extrusions should be in pristine condition and free of scratches and gouges. Streaks may be apparent due to metallurgical inconsistencies in some alloys, but these are not the norm (Figure 6.2).

When delivered from the extrusion mill the extruded lengths should be properly wrapped and protected. The surfaces should be maintained this way throughout the processing.

Large tubes of heavy thickness need to be carefully handled to avoid scratches, dings, and mars. Handling at the mill can be difficult, and since most tubes are provided in bulk order, they often can be streaked and marred unless the order specifies them to be free of unwanted scratches and mars. This may be difficult on small orders. One may be forced to use the available stock and post finish to remove the scratches and surface mars.

FIGURE 6.2 Streaks apparent in aluminum extrusion.

As the material moves through the fabrication process, the underside of sheets and plates can become scratched and marred from passing through equipment and over fabrication tables. If this surface is expected to be visible, it must be protected from damage. This goes for extrusions as well as tube forms.

If the aluminum material is anodized, the surface is more durable than bare aluminum, but it too will scratch and, if scratched, there is no way to reverse it unless you go through the expense of refinishing the surface, removing the anodic film and re-anodizing the part. Removing the anodic film and re-anodizing is possible but the result will be slightly different in color tone. The surface will be altered enough that matching the original appearance in all lighting conditions will not be possible. Anodizing adds cost to the aluminum so the anodized aluminum surfaces should be adequately protected during the fabrication process.

Water should not be allowed to come in contact for any length of time with aluminum, whether mill finish or anodized. It will stain the surface. Storage should be always in a dry, enclosed space.

Aluminum wrought products should be at all times protected from moisture. Never allow moisture to enter between stacks of sheet or plate material. Dark stains will develop on the surface and stains of this type are tenacious and difficult to remove (Figure 6.3).

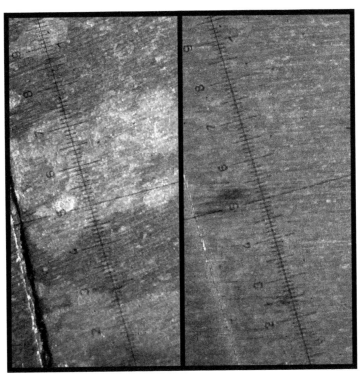

FIGURE 6.3 Stains on the aluminum surface: microscopic images of white and black stains from moisture.

Even if the aluminum is to be processed by anodizing, the etching process will not take out the stain completely. Large fabricated parts stored outdoors should be covered and air should be allowed to flow through the covering to keep them dry. Protecting them with wraps that can absorb moisture can cause staining to the aluminum surface if they are allowed to remain wet.

Mishandling anodized sheet, in particular the thicker anodized coatings, can create microcracks across the surface. The thickened oxide is not as flexible as the core aluminum and if the sheet is mishandled microcracking will occur and this will appear in certain lighting conditions as hairline "spider cracking." Differential heating of the anodized aluminum sheet or allowing the surface to get very hot will create these same small cracks on the surface, as the base aluminum material expands more than the oxide coating.

Cutting

Aluminum can be cut easily with a shear, saw, water jet, high-speed machining mill, an industrial router, and to a limited degree with a laser and a plasma cutter (Figure 6.4).

FIGURE 6.4 Image of a dual saw cutting aluminum extrusion.

For all the cutting processes, aluminum has its own peculiar responses to the tool being used to remove material in order to arrive at the final shape or form. Aluminum's elasticity and low melting point coupled with the metal's ability to move heat away rapidly can make it gummy, and it may adhere to cutting tools. Its high reflectivity and rapid oxidation can affect the way it responds to energy sources, such as a laser, where more energy is demanded for cutting thin sheet. The opposite is true with mechanical processes. The low yield strength of aluminum reduces the energy needed in mechanical cutting processes, such as shearing and the sawing behavior of waterjet cutting. For each cutting method, an understanding of the behavior of aluminum is required.

Shearing

Shearing is a process suited for cutting sheet material, solid bar, solid rod, angles, and wire. Shearing is a method where two blades are brought together from opposing sides of the metal. Shearing and blanking are common methods used to cut straight edges and polygon shapes out of aluminum sheet metal. Shearing is a cold working mechanical cutting process.

The shearing blades have a sharp edge and address the metal with force and speed. Aluminum does not require significant energy to cut but alignment of the shear blades is important to prevent plastic deformation. When clearance and alignment is off, the cutting action becomes more of a tearing action.

Shearing by means of a guillotine style machine is commonly performed on aluminum sheet of thicknesses less than 4.75 mm (0.187 in.). This form of shearing is a straight line cut. It can be 90° off an edge or set at an angle but the cut is always a straight line.

Thin aluminum can also be shaped with slitting, which is done with a pair of wheel shaped blades. As the sheet or coil of metal passes through the blades, the metal is cleanly cut. This also is a straight line cutting operation when used on coils (Figure 6.5).

There are processes that can cut curves using a set of slitting dies on a rotary tool. These are used for cutting disc shapes or simple curves. This operation is similar to opening a can but utilizing more robust industrial equipment.

The punching process used to create holes in sheets is also a method of shearing. This is discussed in a later section. For shearing bar, rod, angle, and wire aluminum forms, a similar, dual blade guillotine style cutting process is used. Guillotine shears can be hydraulically driven or mechanically driven. The metal sheet is set against a square edge and slipped under the blades until it strikes a back gauge. The back gauge precisely measures the distance from the blades so a 90° cut can occur. A clamping system is brought down to hold the metal firmly in position. The blades on a guillotine shear slice the metal sheet in a manner similar to the way scissors cut paper. A crack is initiated on one end and the metal is sliced as the fracture propagates along the force of the blade.

Shearing anodized aluminum can leave small microcracks along the cut edge. As the metal yields at the shear blade the stretching at the localized edge fractures the oxide film. With anodized aluminum and some of the softer alloys of aluminum, the clamping system can impart visible

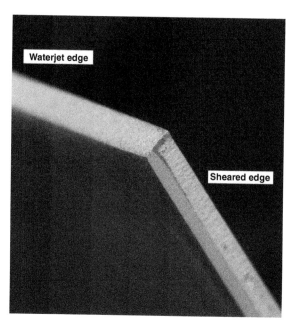

FIGURE 6.5 Comparison of waterjet-cut edge and sheared edge in 3 mm sheet.

depression marks into the metal. It is recommended to protect the top of the metal by placing a thin piece of cardboard between the clamps and the metal surface.

Saw Cutting

Saw cutting is a method of cutting aluminum plate, tube, pipe, extrusion, bar, and rod. It is not typically used on sheet aluminum because the sheet lacks rigidity and catches on the saw blade. Aluminum cuts easily with the correct saw blade and speed. Hardened steel circular blades or steel band saws designed for cutting nonferrous materials are methods used in most fabrication facilities to cut aluminum of thicknesses greater than 3 mm.

Because of the softness of the metal, low melting point and the galling tendency, abrasive cutting and sawing can become a challenge as the blade heats up or if the wrong blade type is used. The metal melts and gums up the abrasive wheel or teeth in a saw blade when the speed is slow or when the blade gets hot. This can be dangerous as the abrasive blade heats up and can explode.

High-speed band saws and high-speed cutoff saws work well for extrusions, tubing, pipe and bar and other three-dimensional forms. Circular saw blade tips should be for nonferrous material. Typically, they are carbide tipped, usually with a titanium carbon nitride. Titanium aluminum

FIGURE 6.6 Assembling large aluminum forms in the factory.

nitride and aluminum chromium nitride are also used. The blade meets the material at a very high rate of speed, approximately 4000 rpm, and exits the material at the same speed. The cut should not generate a lot of heat and be free and clear of burrs. Very little cleanup is necessary if the cut is correct. Circular sawing of aluminum is very fast and efficient. The cut is large due to the thickness of the saw blade.

Saw cutting is also used to v-cut aluminum. Unlike v-milling stainless steel, aluminum will not peel off in a clean cut because of the buildup of heat. Saw cutting, with the saw action at a slight angle and passing the aluminum laterally, can create the v-cut. This is best utilized on thick aluminum sheet or plate. Employ careful control to avoid binding the blade as the aluminum is passed under the rapidly moving saw (Figure 6.7).

When v-cutting to produce a fold, exercise caution; you only have one opportunity to get it right. The issue is potential cracking when the bend is performed. If too deep, the remaining aluminum can be too weak. Flattening and rebending the cut corner can develop a crack (Figure 6.8).

FIGURE 6.7 V-cut corner on the Modern Art Museum.

Handling and Storage 243

FIGURE 6.8 V-cut heavy aluminum plate.

Saw cutting of extrusions and tubes can be performed with a band saw. The band saw must be able to move at high speed, significantly higher speeds than those used to cut steel. Otherwise the blade will gum up and the cut will be poor. Band sawing aluminum will give a smaller cut.

Waterjet Cutting

Waterjet cutting is a sound method to cut aluminum, particularly suited for thick aluminum plate. A computer numerical control (CNC)—directed waterjet can cut intricate shapes and forms in aluminum. Waterjets cut with a high-pressure stream of water with fine garnet particles fed into the water. The water stream is extremely fine and under tremendous pressure, approaching 690 MPa (100,000 psi), and can slice through thick plates. It leaves a frosted edge where the garnet abrasive in the water creates a kerf pattern. If the edges are to be welded this kerf will need to be treated to remove any garnet and minute contaminates that might have been embedded into the aluminum (Figure 6.9).

When cutting with water, the high-pressure stream must pierce through the sheet or plate by subjecting the aluminum to a concentrated stream on a point. This drills a hole through the metal and the stream then moves as described in the direction determined by the control device of the

FIGURE 6.9   Waterjet cutting of thick aluminum plate, by Jan Hendrix.

waterjet. The garnet in the powerful water stream basically saws the aluminum as it moves. There is no heat generated in the waterjet process to cause changes in the alloy, oxidize the edge, or change the local temper of the material.

Waterjet streams produce a kerf along the edge. Depending on the speed of the jet stream as it moves horizontally across the sheet, the kerf is slanted slightly in the direction of the movement of the water stream. For thick plates of aluminum, the kerf is visible and appears as a frosted, grainy surface on the edge with a series of lines.

Waterjet cutting of aluminum is an effective way of producing very accurate and efficient cuts in aluminum sheet and plate. Very detailed shapes can be accomplished. For thin sheet, care should be exercised as the energy from the waterjet attempts to pierce the metal. Bending and distortion can occur. If not supported, small cuts can be tilted out of plane and ruin a piece (Figure 6.10).

FIGURE 6.10 Waterjet-cut and painted aluminum plates. Designed by Kem Studios.

With plates of aluminum, the edge can become contaminated from embedded garnet. Oxides form as new aluminum becomes exposed to air and moisture from the waterjet. The garnet contamination can interfere with welding if used as a preparatory step. Filing or preparing the edge for welding usually alleviates the contamination.

FIGURE 6.11   Thin aluminum art cut with waterjet, by Lee Boroson. Courtesy of the artist.

Anodized aluminum can be cut. There can be some frosting along the edge as tiny cracks are generated in the less flexible thickened oxide coating when the cutting jet of water passes through the aluminum (Figure 6.11).

Machining

High-speed machining is used extensively on both plate and large three-dimensional forms of aluminum. All alloys can be effectively machined; however, harder tempers machine better than softer tempers, and harder alloys of A97xxx respond well to machining. Machining aluminum requires a different set of controls in both the approach and the tooling. Aluminum is more free-cutting than other metals but requires attention to thermal stability and controlling the dissipation of heat. The use of coolants to flush the parts is common to help dissipate the heat from tool speeds operating at 12,000 rpm.

The biggest challenge is the tool itself. Aluminum is susceptible to adhering onto the tool and dulling the cutting blade. Many cutting tools are coated with titanium nitride or titanium carbide, but these should not be used on aluminum. Aluminum will cut better and the tool will last longer if it is uncoated. Use a coarse-grain carbide rather than the finer submicron grade carbide tools. These will not react with the aluminum and are less susceptible to having the aluminum adhere to the cutting tool.

The machine surface of aluminum is very reflective. The edge is created by small slices taken rapidly from the surface of the aluminum as the milling bit carves away the material. The result is a series of shiny cuts. Routing bits act in a similar manner to milling bits. The edges are shiny but sharp. The sharp edge will require dressing down with a file for many architectural and art creations (Figure 6.12).

Aluminum is well suited for the machining of complex shapes. The process is rapid. Five-axis milling can produce very complex shapes and forms. The forms are developed on three-dimensional computer-aided manufacture programs and then translated directly to the mill operation. Tool paths and finish shapes can be examined prior to cutting metal (Figure 6.13).

Routing

Using industrial routers is another way of cutting aluminum. A machining mill is more robust and much faster than a router, but if simple shapes and edges are to be cut, a router can serve the purpose. However, it is important to have the router very rigidly set. Never use a hand-held router and remove only small amounts of metal at time.

FIGURE 6.12 Milled/machined edges on aluminum. Designed by Jan Hendrix.

248   Chapter 6   Fabrication

FIGURE 6.13   Milled/machined aluminum plate. Designed by Gensler Architects.

The bit should be a very sharp carbide bit and care should be taken to keep the aluminum from collecting on the tool. If extensive cutting is needed, consider a machining mill or cut the piece out on a band saw.

Laser Cutting

Laser cutting of metal was introduced in 1967. These early lasers used a stimulated carbon dioxide gas to create a concentrated high-energy beam of coherent light. They were called CO_2 lasers. Laser cutting uses sublimation to cut aluminum. For laser cutting aluminum, oxygen is preferred to aid in the sublimation of the aluminum surface. Oxygen stimulates the cut and removes the sublimated aluminum particles. When focused on the metal surface the metal would absorb the light energy and heat up rapidly to the point it would melt. Another high-velocity stream of gas would blow the molten metal out and you would be left with a narrow cut called a kerf through the metal. The path of the light beam is controlled by an *x-y* CNC table.

Laser cutting depends on light absorption, thus energy absorption. In order to effectively cut aluminum, the electromagnetic energy of the laser light must be transformed into thermal energy within the aluminum. The amount of thermal energy is dependent on the way the metal absorbs the light. Most metals absorb only a portion of the laser light energy hitting the surface. It is this energy that is absorbed by the metal that allows it to be cut.

Light behaves as a photon particle and an electromagnetic wave. It is the electromagnetic wave that will interact with the electrons inside the metal, making them bounce around and repel other electrons, increasing their energy. These electrons become excited by the electric field and this causes them to become agitated and interact with the crystal structure of the metal. This agitation creates heat energy. The coarser the surface, the more light can be trapped or deflected locally on the surface. This further enhances the absorption of the energy and rapidly heats the surface.

The hardness of a material is immaterial to a laser. The amount and type of energy absorbed is critical. Aluminum reflects over 90% of the light wave. Thus, only a small portion of the light is absorbed. A significant amount of energy at a low wavelength is necessary to cut aluminum with a laser. The crystal makeup of aluminum and the wavelength reflectivity of the metal causes much of the light energy to be reflected away. This can corrupt some lasers by reflecting the energy back to the source. Even covering the surface with a nonreflective material won't help since the molten aluminum is very reflective. Lasers are less effective on metals that reflect light or conduct heat. That is why for decades copper and aluminum were difficult to cut for all but the most powerful lasers.

The CO_2 laser is still in use today; however, it is not typically used on aluminum because of concerns about the laser reflecting back to the machine and damaging it. The CO_2 laser delivers energy to the metal while positioned directly above the sheet.

There are solid-state lasers that cut by using light that has been amplified by means of rare-earth doped crystals. These are called YAG lasers or nd:YAG lasers. The letters stand for neodymium doped yttrium aluminum garnet. These are very powerful lasers that can effectively cut thin aluminum.

The fiber laser operates by solid state and delivers the energy via a fiber. Fiber lasers use solid-state diodes and optical fiber that has been doped with rare-earth elements, such as yttrium and neodymium. The energy source delivers the powerful beam of light through the fiber cable, thus avoiding reflections back into the system. The fiber laser is 100 times stronger than the CO_2 laser.

Even with the power of a fiber laser, the thickness of the aluminum is limited in most instances to sheet thicknesses, particularly when the edge cut is important. Aluminum in plate thicknesses that need to be cut should consider waterjet cutting, saw cutting, or machine cutting. Laser and plasma methods may not be appropriate choices (Figure 6.14).

Fiber lasers are more efficient in cutting aluminum than the CO_2 lasers. Once a certain power density threshold is reached, the aluminum is melted and blown away, making the cut. The higher the purity of the alloy, the more difficult it is to laser cut. High-powered lasers can generate significant heat and warp the aluminum sheet, particularly if there are a lot of small holes to be cut in the aluminum. Like waterjet cutting, a hole must first be "drilled" into the metal before the energy beam moves along the path. This concentrates heat in a set point as the metal is heated and blown away by the jet of gas (Figure 6.15).

FIGURE 6.14 Fiber laser cut edge.

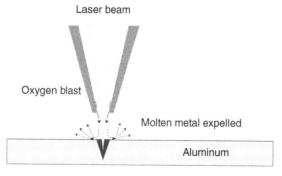

FIGURE 6.15 Drawing of laser drilling.

Thick aluminum will not cut well by laser. The maximum thickness with very high-powered lasers is approximately 8 mm, depending on the alloy. The edge of the cut may require redress due to the melting of the aluminum. Test run parts first to see if the burr left along the cut edge is acceptable or if filing and cleaning can be achieved to satisfactory results.

Slats are used to hold the aluminum as it is being cut. The slats are also cut and can deposit molten steel to the back side of the aluminum as it is being cut by the laser beam. The aluminum thickness that can be cut depends on the power and efficiency of the laser. The edge can be an aesthetic concern, as it is with plasma (Figure 6.16).

Plasma

Plasma cutting of metal involves electrically heated channel of ionized gas. The material being cut must be electrically conductive. Plasma cutting of aluminum is often overlooked due to past practices. The older plasma cutting machines may lack the controls to achieve a good cut in aluminum sheet and plate material. Technological advances in plasma cutting have greatly improved the ability to use plasma for cutting aluminum sheet and plate. Using inert gases to shield the plasma cut edge aids in reducing oxidation along the cut edge of aluminum. Additionally, better speed controls reduce the buildup of heat as the plasma stream moves over the surface.

Early plasma-cutting devices had trouble producing an acceptable cut edge when cutting aluminum. Modern equipment and techniques that use inert gasses to shield the cut and high-speed systems to reduce thermal build can produce a decent edge when using CNC plasma. Cutting aluminum alloys up to 5 mm thickness are possible. Contamination from the steel slats that hold the aluminum during the cutting operation can be a concern. The slat gets cut during the process and can deposit molten metal into the back side along the edge of the cut.

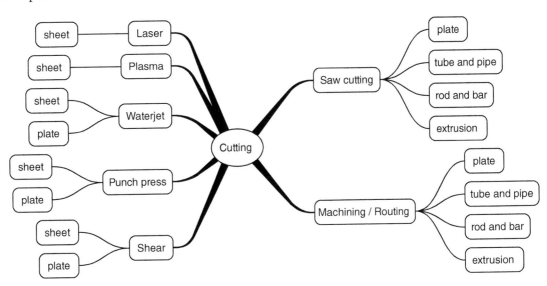

252 Chapter 6   Fabrication

FIGURE 6.16   Fiber laser cut coil-coated aluminum. Designed for Cornell University by Morphosis.

Punching/Perforating

Punch press operations on aluminum are cold working mechanical cutting processes. The energy required is lower than what is needed to pierce many other metals. It is important to use proper clearances and settings for tools to prevent aluminum from grabbing the punch sides and creating ragged holes.

Punch presses are common in the metal fabrication industry. Turret punches, CNC-controlled, are common machines used to punch and perforate metal sheet. The punch operation involves the careful and accurate coordination of four tools: punch, stripper, die, and tool holder. Each of these is engineered to be rapidly deployed in creating accurate and precise holes and piercings in sheet metal. The punch is a hardened steel tool with polished sides, engineered for the operation of piercing or forming metal sheet. The stripper holds the metal sheet and the punch in the correct position for the piercing to occur. The die is below the piece and is accurately positioned in alignment with the die by the CNC-controlled turret. The tool holder holds the die in alignment, both vertically and horizontally.

The critical part starts with the design of the die and punch. The necessary clearances must be designed for the specific metal type and thickness. For aluminum the clearance is from 16% to 19% of the metal thickness being pierced. As with shearing, the piecing creates a fracture at the top side of the cut and at the bottom side of the cut. Proper clearance aligns these cuts so when the piecing is complete, the two cuts meet. When the alignment is off or the clearance is too tight or loose, the metal shears and rips as the force of the tool pushes through. This will add wear to the tool and produce a poor edge on the perforation.

Fracture around the holes on the surface of the metal occurs when perforating anodized aluminum. The hard oxide does not flex as well as the base metal. Punching and perforating anodized aluminum produces a light frosted appearance around the edge of the pierced hole (Figure 6.17).

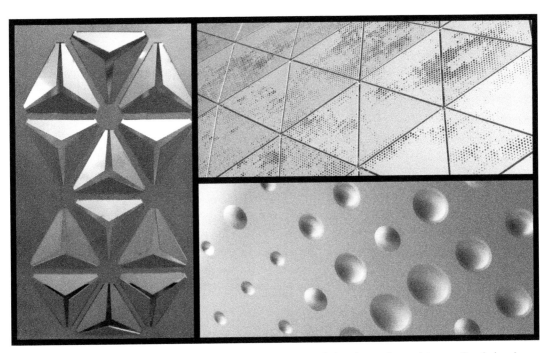

FIGURE 6.17 Custom embossed and perforated surfaces. Natural aluminum, frost white anodized aluminum, and champagne anodized aluminum.

Consider punching the sheet or plate first, prior to anodizing. This will also allow the anodic coating to be applied along the edge of the hole.

Punching can warp the sheet material as uneven stresses are inducted. With aluminum, the amount of warping can be minimized with the appropriate clearance arrangement for the metal thickness. If warping occurs, you can flatten the sheet by rolling through a set of rolls and overstressing the sheets.

There are size limitations on the punch press operations. The machine itself usually has width limitations. They are usually designed for the available sheet widths produced at the Mill for sheet forms of the metal. These usually range from 1220 (48 in.) to 1524 mm (60 in.). Lengths are variable. Tables are designed for a standard sheet length, but a skilled operator can reposition the clamps to do long lengths.

The thickness is constrained by the clamps that hold the sheet as well as the tonnage of the press itself. For most operations, 4.75 mm (0.187 in.) is the thickness limitation that can be punched because of the clamps and because of the tonnage needed to punch thicker material.

Chemical Milling

Another process that is not really a cutting process but more of a surface reduction and design technique is chemical milling. Aluminum can be chemically milled by immersion in strong alkali and selectively dissolving metal. Chemical milling is a reductive process where a masking of vinyl is applied over the aluminum surface and the sheet or part is immersed in a chemical bath that dissolves the exposed aluminum. If the aluminum remains in the etching solution, it will eventually dissolve and perforate.

Chemical etching had its start during the Renaissance when decorative etching was used on steel armor. Today it is used more for creating precision parts and reduction of weight for the aerospace industry. Decorative, intricate designs are possible in sheets of aluminum where the etched region has a frosted reflective appearance and the raised region can have a mirror polish or satin finish. The draft angle is a constant with etching, and this can be layered by repeated masking. The finish work can receive paint or be anodized to create a matte surface finish (Figure 6.18).

FORMING

Forming is a broad term used to describe various methods of plastic deformation of metal wrought forms. Forming is distinguished from forging in that forging of aluminum uses heated dies and heated metal. Superplastic forming of aluminum falls into both camps since that process uses heated dies as well.

Press Brake Forming

Press brakes are used to form aluminum into various linear shapes from sheet and plate material. The process involves using pressure applied to a punch via a ram and pushing the metal into a die.

FIGURE 6.18 Chemical milling of aluminum.

The metal sheet is placed over the opening of a die and the punch is brought down into contact with the metal and as pressure is applied, the metal moves into the die. Press brake forming is performed on small run, custom applications where simple bends and combinations of bends are produced. The process can be more economical than extruding aluminum, roll forming, or press forming aluminum in a stamping press, unless the quantities are significant enough to warrant the expense of the dies for these processes (Figure 6.19).

For most tempers and alloys, aluminum will shape adequately without a significant power transfer from the ram of the press brake. The process of press brake forming creates uneven stresses through the thickness of the metal at the point of the bend. The side facing the punch typically undergoes compression as the metal folds upward while the side facing the die is stretched and undergoes tension.

Sheet thicknesses of 6 mm (0.25 in.) or less are typically press brake formed. As the thickness goes up the tonnage needed to form the metal increases, and difficulties in achieving tolerances also increases as thicknesses exceed 6 mm. Thicker aluminum can be formed; it just requires different measures, not least of which are handling and tonnage.

FIGURE 6.19 Fabricated aluminum cornice and fascia, brake-formed from sheet.

Two additional constraints with press brake forming are bed length and shut height, the distance from the ram to the bed when the die bottoms out. Most press brakes are designed and constructed to meet the requirements of sheet and plate stock, which is typically provided in 3.6 m (12.0 ft.) or

3.0 m (10.0 ft.). There are press brakes that are as long as 9 m (30 ft.) used for making long length panels and other shapes, but they are not common.

The longer the press brake the more compensation is needed to adjust the center of the bed. Most press brakes apply the pressure to the ram from the ends. There can be a very slight difference of pressure applied to the center and this is made up with fine adjustment to the center of the die. Additionally, the dies are made of hardened steel and therefore are very heavy. Storing the dies and placing them into the press brake can be a difficult and expensive set up process. The longer the die, the heavier and more logistical the challenge can be.

In sheet metal fabrication, the press brake is a key piece of equipment. Skilled operators learn the idiosyncrasies of a particular piece of equipment and how to adapt to the changes in material behavior to achieve adequate results.

One of the important pieces of information that a skilled operator must understand is the characteristic described as "springback." *Springback* is the term used to describe a metal's tendency to return to its original form. Most sheet metals exhibit some level of springback. Aluminum is no exception. The amount of springback is dependent on the yield strength of the metal, the thickness of the metal, and the radius of the bend.

The unbalanced tension and compression stresses induced into the metal when it is formed causes springback. To compensate for springback, a skilled operator uses a technique of overbending where, depending on the mechanical properties, some trial and error and general knowledge of the amount of overbending is determined.

From an art and architectural viewpoint, springback is important if several pieces are to be placed together, end to end, or constructed to a level of accuracy that could pose a negative aesthetic if not properly controlled.

Often it is taken for granted that those who are skilled with the understanding of press brake forming will compensate for the springback in order to arrive at a consistent finish product. All metals form differently due to the mechanical properties each metal possesses. When sheet and plate material are produced, there can be differences in the metal even if they are produced in the same coil or at the same time.

Modern equipment and die fabrication design have adapted techniques to aid the operator of the press brake to arrive at a consistent and predictable bend in the metal.

For example, dies are manufactured to compensate for the springback in the metal by special adaptions. Many dies used for simple bends and forming in the sheet metal fabrication facility use air dies. This simply means that as the metal is pushed into the die by the punch, there is air below the metal (Figure 6.20).

You do not want the punch and die to come into contact. The metal is not bottomed out in the die during the press operation. A die may be made with an opening that is 85°. This can allow for as much as 5° of overbending to achieve a 90° angle. For aluminum, typical alloys used in art and architecture will have 1.5–2° of springback, depending on the yield strength and the angle. Tempers of H32, common in the A95xxx alloys used in architecture, may need greater allowances, up to 3°. Higher yield strength alloys, such as A92xxx and A97xxx, will exhibit a greater amount of springback, between 2° and 3°.

258　Chapter 6　Fabrication

FIGURE 6.20　Perforated and folded aluminum, natural finish. Designed for the Nerman Museum of Contemporary Art by Kyu Sung Woo Architects.

A skilled operator will determine the radius of the finished angle. One technique used is to determine the springback factor, S_F, and multiply this by the desired radius to determine what radius is needed in the press brake operation. So, for a 2 mm thickness aluminum sheet, we may want a 2 mm radius (Figure 6.21).

The springback gactor is determined by dividing the angle desired by the angle needed for the material springback. So, for aluminum, if we use a springback of 2° and we want to end with a 45° angle, then we need to press the metal to an angle of $45 - 2 = 43°$.

$$S_F = \text{Desired angle/Press angle} = 45/43 = 1.04$$

$$\text{Radius desired} = \text{Radius pressed}/S_F = 2/1.04 = 1.93\,\text{mm}$$

The operator would press the metal into the die and measure the angle radius that results. The operator would press the aluminum into the die until a radius of approximately 1.93 mm is achieved. When the pressure is released, the aluminum would springback to the 2 mm desired.

FIGURE 6.21 Folded aluminum plates, designed for Emerson College by Morphosis.

Certain die designs used by press brake operations can facilitate the overbending without bottoming the metal into the die. There are dies for forming corrugated panels, dies for curving and creating cylindrical forms, roller, and sweeping dies for producing tight folds (Figure 6.22).

There are rubber pad forming dies, where a conventional punch is used but the die is a certain durometer rubber held in a steel retainer. This die acts as a diaphragm and produces forces similar to hydraulic action. These rubber pad dies allow special complex shapes to be formed that cannot be accomplished with standard die and punch setups.

Urethane rubber is typically the rubber that is used due to its durability and the fact that it resists compression. Urethane, under pressure, will change its shape and this characteristic is what pushes the metal around the punch in an even and uniform manner. When the pressure is released, the rubber pad returns to its original configuration.

Urethane dies are for short production runs. They also require more tonnage as the metal is pushed into the compressed die. They eventually wear out and lose some of their strength. Urethane rubber comes in various tensile strength grades from 18 to 76 MPa (2.6–11 ksi). Typical durometers used are 75D, 80A, 90A, and 95A. The 75D is considered extra hard and is frequently considered

FIGURE 6.22 Formed and post painted aluminum for garage façade, designed by IwamotoScott Architecture.

for the toughness required in this type of forming. Solid blocks require as much as three times the tonnage as an air die. Often an airspace is set below the die to give the urethane rubber some space to move into. Many urethane dies have an extruded space running the length internal to the form. This helps to keep the tonnage down. The shape deflects under load and the urethane is pushed up to engage with the metal.

One of the best uses of urethane rubber pad brake forming is when the radius of the bend is significant. An experienced press brake operation can form accurately curved sections using rubber pad forming. The urethane pad can shape the metal evenly and not impart marks or indentations onto the metal.

Roll Forming of Aluminum

All alloys of aluminum can be roll formed. Roll forming, however, requires a quantity of parts of similar cross section. Aluminum sheet is an excellent material for shaping with stages of rolls.

Whether prefinished, embossed, or mill surface, shaping with properly designed rolls is an efficient way of arriving at large, long length surfaces using aluminum sheet. Roll forming operates in stages of polished or chromed rolls. Each subsequent roll shapes the metal a degree at a time. Proper roll forming stations are long, mechanically driven machines that receive the metal sheet or coil and form it by passing it through matching rolls set for the thickness of the material.

Roll forming is a viable way of adding section to a thin aluminum sheet. The patterns are always linear along the length of the sheet. This process is used extensively in metal roofing applications where long, unbroken panels are needed to cover a surface. The edges are designed to interlock and keep moisture out while still allowing for the thermal movement of aluminum.

Roll forming dies and equipment are expensive but once the configuration of the form is created and the rolls are properly aligned, the roll forming operation is inexpensive, fast, and productive. The key to roll forming aluminum is to control the introduction of stress to the aluminum sheet. Begin with stretcher leveled aluminum sheet. There are several types of in-line levelers that receive the ribbon as the coil is unwound. The stress is overcome by pushing the sheet up and down through a series of rolls, which effectively distributes internal stresses evenly across the sheet before it enters the roll forming station.

The roll forming station is a series of rolls that are designed to incrementally bend the aluminum. If the bend is slight, then generated stresses are small as the metal undergoes localized stretching between each roll. The roll forming stations can be relatively short for simple bends or quite long for major bending. There can be over a dozen rolls in sequence, each producing a minor additional bend into the sheet. If the bend is significant between roll stations, stress will be induced into the sheet. This stress can manifest itself as a buckle in the sheet or in-and-out undulation commonly referred to as roll forming.

Stress can also develop if the clearance is too tight between the mating dies. This stretches the metal locally and causes ripples in the metal. Aluminum is ductile and soft compared to the machined rolls. If the rolls are too tight, they can rip the metal or squeeze the edge, imparting a marking running the length of the material.

Stretch Forming of Aluminum

All aluminum alloys can be stretch formed. This is a cold forming technique. In this process an aluminum shape is pulled over a solid form or die to the point where the metal yields and elongates slightly as it acquires its new shape. A simple form of stretch forming is used by extrusion companies to pull the twist out of an extruded shape that has undergone shaping from differential stress induced as the part cooled. The ends of the extrusion are firmly held in mechanical jaws and pulled. The extrusion stretches just slightly but enough to redistribute the stress (Figure 6.23).

In more complex stretch forming operations, the aluminum formed shape is held at the ends as a large die is pressed into the shape. The metal form is wrapped over the die stretching to the curvature of the large die. There are numerous custom shapes that involve curvature produced by stretch forming operations.

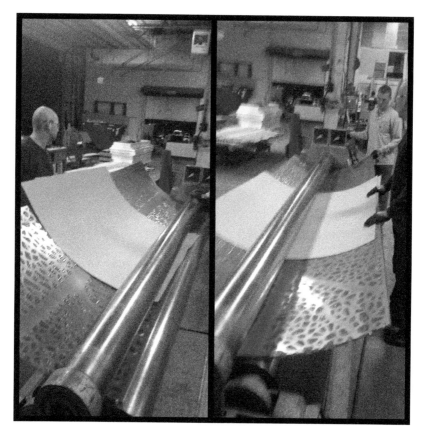

FIGURE 6.23 Plate rolling large aluminum plates.

One unique system uses stretch forming to shape aluminum plates into dual curvatures by grabbing the ends of the plates and pulling slightly while a die made of a matrix of actuators forms the dual curve into the plate. The dual curve is induced accurately into the aluminum sheet.

Hydroforming

Hydroforming is a cold forming process that utilizes hydraulic pressure to push the metal into a die. More often used in the automotive industry on ductile aluminum sheets, this method of forming can impart stiffness by pushing the sheet under hydraulic pressure into a die. The forming process can also shape a tube form into a cylindrical die by using liquid under high pressure. This cold forming process uses oil, often held within a bladder to contain the oil. The pressures are significant as they push the metal into or over a steel die. The finish on the aluminum surface after forming is smooth, unlike the graininess developed on the surface of aluminum when stamped or hot formed.

Spinning

Aluminum is readily formed into symmetrical shapes by spinning. Initially the aluminum is cut into a circular blank. Blanks can be as large as 2 m and as thick as 6.4 mm (0.25 in.) and spun at room temperature. Even larger blanks have been spun but these require heating and significantly more advanced equipment (Figure 6.24).

In architecture and art, the alloys most often spun are the softer alloy forms of A91100, A93003, A95052, and A95086. If the forming is to be severe, special fully annealed blanks are considered (Figure 6.25).

FIGURE 6.24 Spun mill finish aluminum balloon shapes on aluminum rods, by William Zahner.

FIGURE 6.25 Spun mill finish aluminum clock form, by Ron Fischer.

Spinning can be very precise as the metal is formed into or over a die. The process begins with clamping a circular disc to a mandrel. The mandrel is rotated as a tool called a roller is pressed against the blank where it is supported by the mandrel. The material rolls over the mandrel a small section at a time. The aluminum disc is progressively formed at each pass of the tool. Once the sheet is formed over the mandrel, a final pass can be performed to produce a fine, polished surface with a tight tolerance over the mandrel forms. Hemispheres, dish forms, and bullet forms are spun from aluminum blanks (Figure 6.26).

Stamping

Stamping is the general term given to describe the operation of pressing a die into a sheet of metal. It is a cold forming operation usually limited to shallow dies. We think of stamping adding section and detail to a ductile aluminum sheet like the stamped metal ceilings made of thin sheet metal. Aluminum is ideal for many stamping operations. The dies must be designed in such a way to alloy metal to flow without necking and cracking. Tight corners or 90° bends are

FIGURE 6.26   Spun light fixture forms in mill finish aluminum, by Ron Fischer.

discouraged. The metal will be restricted and rip as it tries to move into tight or sharp corners (Figure 6.27).

Stamping can be performed on large sheets of aluminum. Limits, other than the design being imparted, are the size and tonnage of the press used to push the die into the metal. There can be a positive and negative mating die, or a rubber pad used to push the metal into the die. The metal needs to move before the dies bottom out or the aluminum will tear (Figure 6.28).

Superplastic Forming

Superplastic forming of aluminum is a forming method used to develop intricate shapes using aluminum sheet. This method is similar to stamping processes but heat and pressure are added.

FIGURE 6.27 Stamped and painted aluminum forms in wall façade for retail.

The alloy and temper for this type of forming are specially prepared to arrive at very fine grains. Large plastic deformation occurs as the hot metal is forced over a heated surfacing tool using gas pressure and vacuum. The expense is in the custom dies needed to arrive at the shape as well as the lower production rate.

Forming 267

FIGURE 6.28 Large stamped aluminum forms for wall façade, postpainted.

Very detailed surfaces can be created by superplastic forming processes. Once the aluminum is removed from the die, the surface is blasted to remove tool lubricant residues and oxides that have formed. This gives the aluminum finish a frosted appearance. This slight tooth aids in subsequent paint applications to the finish surface.

268 Chapter 6 Fabrication

The process is relatively slow and expensive, but the result can be a deeper, more detailed surface than stamping processes (Figure 6.29).

Forging

Aluminum is well suited for forging. Forging is a process like stamping where a metal blank of sheet or thin plate is heated and set onto a heated die. A matching punch is brought down, and the aluminum assumes the shape. Forging is a hot process. The aluminum is set at temperatures approximately 56°C below its solidus[1] temperature.

For aluminum, forging operations are performed using mechanical or hydraulic presses. The equipment is no different from that used on other metals. A lubricant is used to keep aluminum from sticking to the dies.

FIGURE 6.29 Superplastic formed panels.

---

[1] Solidus temperature is the temperature where molten aluminum becomes solid. For aluminum this temperature varies per alloy. For A96061, this temperature would be in the range of 432–482 °C.

FASTENING

Fasteners used in aluminum fabrications, either to join aluminum to aluminum or aluminum to subframing, are often the subject of debate. The concern hinges on galvanic corrosion caused by the two differing metals used. The fastener is usually a metal different from aluminum, such as a stainless steel fastener used in curtain wall assemblies. Since aluminum is more active than most other metals, galvanic corrosion of the aluminum is the subject of concern.

Aluminum bolts, screws, and threaded rods are available in aluminum. The typical alloys are A92024, A96061, A96262, and A97075. These are stronger forms of aluminum, but they can exhibit galvanic reaction to the base metal. The fasteners can be anodized to improve corrosion resistance or treated with a dichromate. But still, aluminum fasteners in aluminum assemblies can be sacrificial to the base aluminum. Figure 6.30 shows an aluminum fastener of alloy type A95xxx in an aluminum plate alloy of A96061. After 40 years the oxidation is more rapid on the fastener as it sacrifices to the plate. Table 6.3 shows relative galvanic relation of the different aluminum alloys.

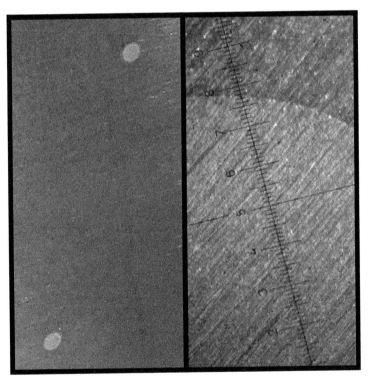

FIGURE 6.30 Aluminum fastener corroding due to slight variation in alloy from plate.

TABLE 6.3 Electromotive force relationship of various alloy categories of aluminum.

More anodic
A92xxx
A97xxx
A95xxx
A91xxx
A93xxx
A96xxx
More cathodic

Another reason aluminum fasteners are not considered is due to the shear strength of aluminum, which is low compared to other metals. Fasteners are often subjected to shear paths as the loads are applied to a surface. Many instances require the fasteners to receive sufficient torque loads when joining parts. The low shear strength of aluminum often forces the engineering need for stronger fasteners.

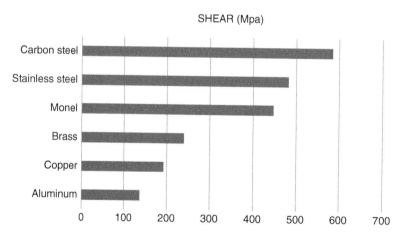

Aluminum extrusions and aluminum panels are used in exterior applications where they are subjected to both significant loads and to exposures where corrosive conditions are present. Therefore, many of these conditions utilize stainless steel fasteners to join aluminum parts. The stainless steel is more noble than the aluminum and the larger mass of the surrounding aluminum as compared to the small fastener mass reduces the corrosion rate significantly. Very little galvanic corrosion occurs when stainless steel is used. Another benefit with the stainless steel fastener is the hardness of the metal in relationship to the aluminum. Screw guides designed into aluminum extrusions do not need pre-tapping to create threads. The stainless steel screw thread will tap the walls as it is engaged.

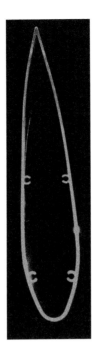

FIGURE 6.31   Image of screw guide.

One drawback is overtorqueing the fastener and stripping the threads. The softness of aluminum and the low yield strength can make it difficult in high torque conditions. Another precaution is that pre-tapping the holes should be performed after anodizing or painting. Anodizing will change the dimension slightly of the hole and painting can fill the threads (Figure 6.31).

Zinc plated steel fasteners, zinc-aluminum plated fasteners will perform adequately in many exposures, but the plating will eventually wear, and certain exposures cause the steel underneath to corrode. The aluminum couple does not afford the same level of galvanic protection as zinc.

SOLDERING AND BRAZING

Brazing

Brazing is the process whereby a metal section is joined to another section by creating a metallurgical bond between them. Aluminum can be brazed, but it is not a simple task. It will work for joints that are not subject to a significant amount of stress.

Brazing aluminum is not suitable for butt or fillet type joints. Lapping joints can be brazed by using the same metal alloy that you are joining. The difficulty arises, however, as you try to heat the brazing metal to the melting point without damaging the host metal being joined. If you have to bring the temperatures to melting, the entire joint can melt. Brazing is performed with an oxy-acetylene torch set. The torch is set rich in acetylene and reduced oxygen. An aluminum brazing flux is employed to aid in creating a sound, brazed joint. For most art and architectural purposes, other techniques are usually considered first.

Soldering

Aluminum in art and architecture is rarely soldered. It is possible but requires skill and the use of special lead and cadmium free solders as well as the correct alloy selection. Alloys of aluminum that have magnesium as an alloying constituent are not able to be soldered unless they are coated with other metals, such as nickel or zinc.

Solderable Alloys
A91xxx
A92xxx
A93xxx
A94xxx
A97xxx

The challenge with aluminum is the oxide on the surface and the low melting point. Special fluxes are needed to remove the oxide and solders containing zinc and tin are used to make the solder joint. The oxide must be thoroughly removed and when the soldering is complete the surface must be flushed and cleaned.

Solder Types Used to Solder Aluminum

98% Zn	2% Al
91% Sn	9% Zn
70% Sn	30% Zn

WELDING

Aluminum is an excellent metal for welding, although, it requires a clear understanding of the characteristics of the metal to achieve success. It is not the same as welding steel. When welding steels, a good weld can be as strong as the steel being joined. This is not the case with aluminum. The weld is always weaker in all but fully annealed aluminum. When considering welding aluminum, the following are critical considerations:

- End use
- Alloy being welded
- Strength requirements of final product

Another difference with welding aluminum is the equipment used to weld. The low melting point of aluminum might lead one to believe that welding currents can be lower, but this is not the case. The thermal conductivity of aluminum is significantly more than for steel, nearly five times as conductive, so the metal wants to pull the heat away and cool down quicker. Because of this the equipment must be capable of operating at higher currents and higher voltage levels. It takes approximately 250 A to weld 6 mm thick aluminum and 350 A to weld 12 mm thick aluminum.

One of the issues with welding aluminum is the incomplete fusion at the beginning of laying down the weld. The higher thermal conductivity of the aluminum pulls the heat away as the surround metal acts as a heat sink. To overcome this, there are hot startup features on modern welding equipment that allow for an increase in energy at the start and this increases the heat input. After starting up the energy drops down to enable the rest of the weld.

These variations in equipment and in technique used to weld other metals are necessary for successfully welding aluminum. Water-cooled torches will allow welding at higher duty cycles with this increase in amperage necessary for welding aluminum. Additionally, spray arc transfer when GMAW (MIG) welding is recommended. Short arc transfer used on steels will not work for aluminum. There is just not enough heat generated to get adequate fusion of the metal. This is not easy when welding thinner aluminum. In this case consider pulse MIG welding equipment. This allows for the current to be reduced and you won't burn through the thinner aluminum.

If GTAW (TIG) welding aluminum, use AC polarity on thin aluminum and DC for thicker aluminum. Increasing the amount of helium will also create a more concentrated heat at the point of weld.

If visual requirements are critical, post anodizing assemblies that have been welded will result in an altered appearance at the weld. The welds will be visible as a slightly different color or sheen. Additionally, preanodized parts that require welding will need to have the oxide produced by the anodizing process removed where the weld will occur. You can expect cracking or crazing of the anodized surface near the area where the heat from welding created expansion stresses in the anodized surface.

Welding aluminum can be achieved by following certain procedures that take into account the high thermal characteristic of aluminum and the oxidation of the surface of the aluminum. Start with a very clean, oxide-free edge or surface. If grinding is used, only dedicated abrasive pads should be used. Stainless steel wire brushes free of any minute steel particles can be used if new and dedicated to the aluminum. Be sure there are no solvents on the surface or in the event of waterjet-cut edges, and clean the edge of all garnet or abrasive residue that may be embedded in the cut edge. It is important to eliminate hydrogen from the weld joint. The presence of hydrogen will increase the weld porosity. As the weld solidifies pores will be generated as the hydrogen escapes. The sources of hydrogen are hydrated aluminum oxide, organic substances, such as oils or greases, moisture from condensation, and even small leaks from water-cooled torches. To eliminate hydrogen, make sure the part is clean and dry. The wire and filler metal used for welding is kept in a heated cabinet.

The wire feed system used on GMAW welding processes on aluminum is more critical than that used on steels. Due mainly to the mechanical properties of aluminum, its low tensile strength, and its softness, aluminum wire is subject to distortion, shaving, and narrowing. Braking systems used on the feed of wire should be at minimum to keep the metal from stretching. Nonmetal tips and guides should be used to keep the wire from losing diameter by shaving off layers as the wire is fed. Planetary drive systems on wire feed operations assist in reducing feed issues on aluminum wire.

Aluminum's high thermal conductivity requires higher amperage and voltage as compared to steel welding. Heat will dissipate quickly with aluminum due to the metal's high thermal coefficient, which is several times that of steel. Additionally, higher travel speeds when applying the weld are needed to keep the heat affected zone to a minimum and thus reduce the development of stress cracking.

There is a finesse to welding aluminum that comes with practice. Unlike steel or stainless steel where you work to "pull" the weld into the gap, with aluminum you "push" the weld. You need to be careful not to dwell long in one spot, particularly with thin aluminum. You can melt or burn through the metal due to the lower melting point.

The weld shape should also be built up, appearing as a dome shape in cross section. The convex shape helps reduce cracking as the metal cools and contracts. Concave shaped welds crack as contraction from cooling increases the stress as the edge draws in. Microcracking is one of the most difficult effects to control while welding aluminum. It is nearly impossible to avoid entirely, but for structural requirements it is important to reduce cracking, even microcracking, to the greatest extent possible.

At the end of the weld run, it is important to reduce the formation of a crater. This will create distortion in the part as the metal shrinks and can create cracks. The technique is to reduce the weld pool just before the arc is closed. Shielding is important for a clean weld with aluminum. Argon shield gas is typically used. It will keep the weld clean and oxide free. For welding the A95xxx alloys of aluminum you should add helium to thwart the development of magnesium oxide (Figure 6.32).

If structural requirements are critical for the welded aluminum assembly, then the engineer must design for the weld. The weld will dictate the strength needed in the part or assembly. Unlike steel, with aluminum the weld is always weaker than the alloy being welded. Depending on the alloy and the tempering of the alloy, the temperatures of welding can have a significant effect on the strength. There are two basic types of aluminum alloy, those that can be heat-treatable and non-heat-treatable alloys.

Heat-Treatable Alloys

A92xxx

A96xxx

A97xxx

Non-Heat-Treatable Alloys
- A91xxx
- A93xxx
- A94xxx
- A95xxx

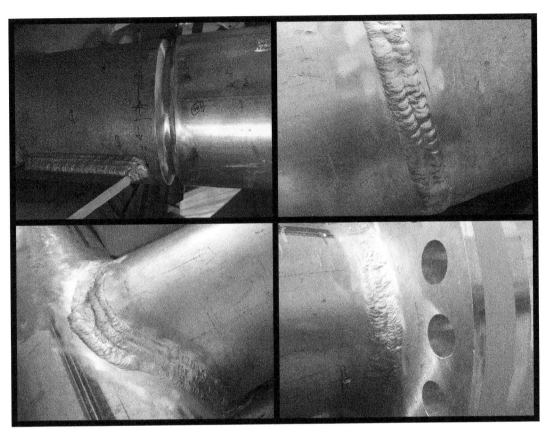

FIGURE 6.32 Weld preparation and weld on thick aluminum tubing.

Welding Heat-Treatable Alloys

For heat-treatable alloys, the last heat treatment usually is in the range of 162–205°C, (325–400°F), so it is important to not overdo the preheating of the metal prior to weld. The preheating also will aid in getting moisture out. So, just before application of the weld, preheat the metal to only 93–110°C, (200–230°F). This will reduce cracking of the material around the weld when it cools and will not anneal the surrounding metal. The weld temperatures take the metal around the weld to over 204°C (400°F), which will anneal the metal and some strength will be lost. Post weld heat treatment can improve the metal around the weld for the heat-treatable alloys (Figure 6.33).

Welding Non-Heat-Treatable Alloys

Non-heat-treatable alloys acquire their strength from cold working. The heat of welding will have an effect on the material around the weld and alter the mechanical properties. For non-heat-treatable

276   Chapter 6   Fabrication

FIGURE 6.33   Welded aluminum plate, by Chuck Von Schmidt.

alloys, preheating is not as critical. The heat-affected zone around the weld returns the alloy to the weak annealed properties.

The A96xxx alloys are prone to hot cracking. Hot cracking occurs as differential shrinking develops. It is critical to choose the correct filler metals and to properly design the weld joint. The joint must be adequately beveled to allow for good base metal dilution with the filler. The beveled edge allows for more filler metal to be placed in the weld (Figure 6.34).

For the non-heat-treatable alloys, overheating can reduce the strength. When preheating, these alloys keep the temperature to no more than 65°C (150°F).

Filler metals for aluminum are the same aluminum alloys used to fuse the material. Use A95356 to improve strength in the weld. Also, alloy A94043 can be used. With alloy A94043 there will be some loss of ductility.

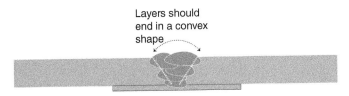

FIGURE 6.34 Joint design is critical when welding aluminum.

Fusion Stud Welding

Aluminum can be stud welded similar to steels. There are a few challenges that aluminum presents that differ from steel. Aluminum conducts heat better than steel. This can create a cold weld by cooling the aluminum down faster. The oxide that rapidly forms on aluminum must be removed subsequent to the stud-welding process. This is usually accomplished by sanding off the surface where the stud will be positioned. The oxide is less conductive and will prevent fusion leading to poor welds.

There are two common methods of stud welding:

- Capacitor discharge stud welding (CD)
- Drawn-arc stud welding

In capacitor discharge stud welding a specially designed aluminum stud is used. There is a small point tip that establishes the proper gap between the stud and the workpiece. It is important that the stud sets perpendicular to the face being welded. Contact is maintained by pressure using a spring in the stud gun. When the electrical current is discharged through the stud, the gaping point melts along with the adjoining faces and the spring drives the stud into the molten pool. The entire weld process occurs in milliseconds.

In the drawn-arc process, a shielding gas is used. The stud is a special design somewhat different from a CD stud. The process also uses a ceramic ferrule that delivers the shielding gas to the area around the weld. The gas is helium or argon or a mix of the two. Helium works well by improved arc ionization and focusing the heat on the weld (Figure 6.35).

Stud welding is an excellent method of attaching stiffeners to the reverse side of aluminum plates. The challenge is to limit the amount of visible show through the material thickness to the face side. The addition of high bond tape and silicone adhesives combined with the stud welding process can often eliminate the reading of the small stud (Figure 6.36).

It is important to check the settings and the visual condition on the face of the part early in the process. Establish a protocol where the settings match with the minimum visible effect on the exposed side of the assembly. Check the tensile strength and perform a bend test by striking the fused studs with a hammer, or better still use a stud pull test device to ensure proper fusion. If using a hammer on a test stud, the stud should bend over before popping off. If the underside of the stud is shiny, then the surface is melting but has cooled too rapidly for fusion to occur. Additionally, there

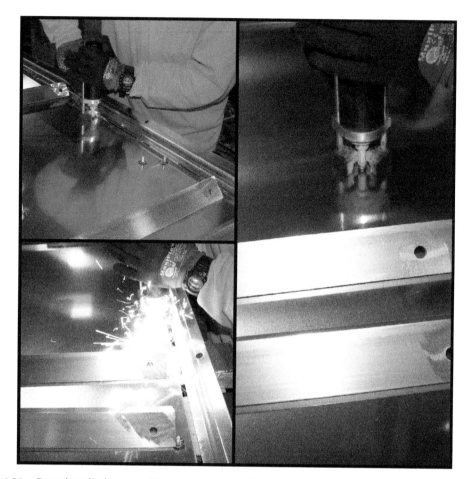

FIGURE 6.35 Capacitor discharge welding on aluminum plate.

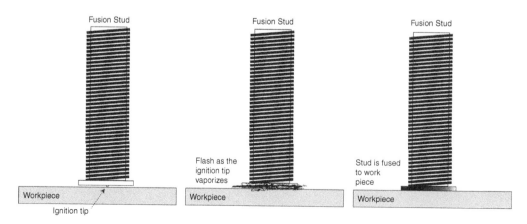

FIGURE 6.36 Stud welding process.

should be 360° of flash around the stud. If this is not the case, then the stud gun is not perpendicular, or something is incorrect in the stud makeup.

These initial tests should be performed on the same alloy of material. Several test welds should be performed each day and a record kept of the settings. Periodically during the operation, run checks again and visually examine the face and stud for inconsistencies. Capacitor discharge machines, like other mechanical devices, will wear out and the capacitor charge will change over time.

Friction Stir Welding

Friction stir welding is a process of joining two plates of metal together by passing a specially designed rapidly spinning tool between them. This process of welding was developed in 1991 and used extensively in the aerospace industry to weld A92xxx and A97xxx alloys. The heat developed by friction melts the metal of both plate edges where they meet the tool. As the tool passes between the two plates the metal just behind the tool is molten and as it solidifies it welds the plates together. The design of the tool keeps the metal contained while spinning at a rapid speed. Friction stir welding does not consume any material as the rapidly rotating tool is inserted into the butted edges of two plates.

The process is not in common use for art and architectural projects, but its uniqueness presents a method of joining two flat plates together without the addition of other metal. The weld joint is stronger than other methods of welding and the heat affected zone is smaller as well. Corrosion resistance is reduced due to the microstructure changes around the weld point.

> Aluminum is an excellent metal for welding, although, it requires a clear understanding of the characteristics of the metal to achieve success. It is not the same as welding steel.
>
> For structural requirements, designing for the weld is paramount.

CASTING

Compared to steels or copper alloys, aluminum is somewhat more difficult to cast. Most alloys have low fluidity. Fluidity is the term used to describe a metal's ability to flow in the molten state. The better the fluidity, the more detailed the casting can be and the thinner the section. Most metals shrink as they cool. Aluminum alloys will undergo shrinkage at a higher rate and will continue to shrink even after solidification as the cast element cools. Good mold designs can accommodate this, but the low strength of aluminum can lead to cracking as stresses develop during solidification.

Foundries with experience in casting aluminum understand the differences in aluminum casting versus casting of other metals. One major advantage is the low melting point of aluminum. Energy use is less, which allows for remelting and reworking.

Sand casting aluminum is one of the more common aluminum casting processes in art and architecture. Aluminum is well suited for sand casting. A broad range of sand types can be used

FIGURE 6.37 Examples of modern casting of aluminum.

with less emphasis on refractory characteristics. Synthetic and natural sands with fine grains can be used, thus enabling a finer surface. The low weight and low melting temperature produce less pressure on the mold.

Aluminum castings are often smaller than castings of other metals. Assemblies of cast aluminum are often bolted together. Welding aluminum weakens the metal around the weld, so you must design for the weld and use thicker sections or internal frames (Figure 6.37).

Pure aluminum is difficult to cast because of the high shrinkage that occurs. Pressure needs to be applied and larger risers are needed to accommodate feeding the metal as shrinkage occurs. Most castings use various alloys of aluminum to achieve particular results.

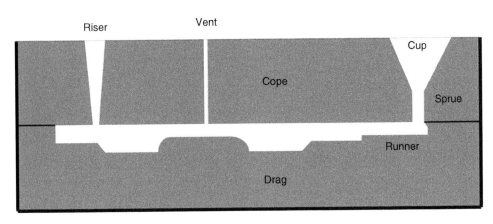

FIGURE 6.38   Basic diagram of a sand cast mold.

With aluminum casting, there are four critical factors that need to be understood (Figure 6.38):

- Fluidity
- Shrinkage
- Dross formation
- Temperature

Fluidity describes how well the molten metal will flow into the mold and fill the void. Good fluidity allows for fine detail and thinner sections to be achieved. As the metal is fed into the mold, it travels to all the reaches of the void.

Shrinkage needs to be understood by the designer of the mold and foundry. As the metal solidifies, shrinkage influences the quality and soundness of the casting. Since aluminum alloys contain different elements, these elements will solidify at different rates.

There are three categories of solidification shrinkage: narrow, medium, and wide. These describe how the metal solidifies as it cools. Narrow shrinkage occurs from the mold wall edges inward and from areas where the mass is less to areas where the mass is greater. Areas where mass is greater are thick regions that remain hotter while the thinner regions cool and solidify more rapidly. Narrow shrinkage range will need to have larger risers to feed metal. A good mold designer will incorporate techniques to direct how solidification occurs. Placing chill bars in areas to cause them to solidify first, or adding tapered sections closer to the riser keep the thermal mass where it will solidify last.

Medium shrinkage alloys are the easiest to work with. Small risers thattake up some shrinkage as the metal solidifies are all that is needed.

The wide range shrinkage alloys are the most difficult to handle. They tend to solidify all at once. Adding risers or tapering the design have little benefit. Keep the section as uniform as possible.

FIGURE 6.39   Sand cast mold: drag portion showing design and raw casting.

Figure 6.38 shows the major components in a sand cast mold. The molten metal is fed into the cup and enters the cavity by traveling through the sprue and runner. As it fills the cavity, air and hot gasses are evacuated by the vents and through the porous mold. As the molten metal fills the cavity it reaches the riser. You want molten metal to enter the riser, and while the metal cools and begins to shrink, the riser supplies metal to the part.

Dross formation is one of the characteristics that can have an effect on the quality of the aluminum surface. Molten aluminum exposed to the air rapidly forms an oxide layer. This oxide is lighter than the metal and floats to the top. The surface next to the drag will have the better quality because it will be free of the oxide. The dross will rise to the top, making the surface next to the cope more porous. The dross is removed by blasting the surface.

With sand casting aluminum, the removal of gases by means of mold porosity is more important than with casting other metals. The hot gases that develop must be driven through the porous mold rather than through the metal. The biggest difficulty with aluminum is achieving the right amount of mold porosity. What can happen when the gasses are directed into the metal is that hydrogen will develop and create a porous finish surface.

Pouring temperature is critical to eliminate hot spots that can form in the casting. The hot spots will affect the metallurgy and surface quality (Figure 6.39).

RAPID PROTOTYPE

There is a lot of discussion about direct part fabrication using methods developed out of the rapid prototype process. Whether low-temperature metals such as aluminum will ever be a viable option as the material component of direct manufacturing or 3D printing is yet to be answered. The difficulty will lie in the metallurgy and the way aluminum oxidizes on exposure. Additionally, fine aluminum powder can be hazardous in an explosive context.

Where significant strides have been made is in molds for casting. The molds are first developed as a solid computer model, printed using rapid prototype machines, then coated with a ceramic to create the dies for casting. Aluminum has such a low melting point that rapidly casting into a basic mold is one of the big advantages of the metal.

Laser sintering is a possibility, but the process so far is limited to small parts not subject to a lot of force. There is still the oxidation concern of the small beads of aluminum being melted together.

CHAPTER 7

Corrosion Characteristics

And I was standing over there rusting for the longest time.

—Tin Woodman, in *The Wizard of Oz*

Aluminum is a very active metal. On the electromotive scale, it is usually found down near the anodic, sacrificial side of the scale. Table 7.3 shows an example of how aluminum is located relative to other metals on the scale. However, this does not tell the full story of aluminum. When aluminum is exposed to oxygen or oxidizing agents, a very thin, strongly adherent film develops. This film changes the corrosion behavior in a fundamental way and dramatically lowers the chemical activity of aluminum.

There are numerous instances in which aluminum has performed remarkably well for decades with no additional coating. Aluminum streetlight support poles and aluminum flagpoles are rarely coated or finished. Yet, they perform well with little more than the occasional rain to remove the road salts and road detritus. They may lose some of their original shine, but they show no signs of significant corrosion.

The oxide film that develops on aluminum surfaces is resistant to oxidizing salts made of nitrates and sulfates and to many oxidizing acids. The oxide is less resistant to halogen acids, such as hydrochloric acid and hydrofluoric acid. These will attack the surface of aluminum and even diluted solutions will etch aluminum.

It is the alloying elements within aluminum that pose the greatest challenges from a corrosion perspective. These elements can form cathodic compounds with a positive electrode potential to the surrounding aluminum. These interstitial elements develop in solid solution and are precipitated within the oxide film, where they can affect the corrosion resistant nature of the film (Figure 7.1).

Silicon, iron, copper, manganese, nickel, and magnesium will combine with aluminum in solid solution. Silicon is a common impurity in aluminum, but the electro-potential is close to aluminum at about 0.05 V. Thus, it does not have a detrimental effect on the corrosion resistance. Iron, another

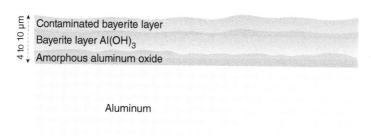

FIGURE 7.1 Cross section of the aluminum surface.

impurity common in aluminum, is insoluble and will precipitate out of solid solution as $FeAl_3$ or $Al_6Fe_2Si_3$. These compounds are more cathodic in relation to the surrounding aluminum and can set the stage for the development of a corrosion cell.

Copper is soluble in molten aluminum to approximately 5%. As the metal cools, copper precipitates out into $CuAl_2$. The precipitation of the copper aluminum compound occurs along the grain boundaries, which will have a detrimental effect on the corrosion resistance of the aluminum alloy (Table 7.1).

All metals experience corrosion to some extent and aluminum is no exception. Corrosion is a broad term that relates to the tendency of metal to succumb to natural forces that seek to return it back to a mineral form with lower energy. Energy is stored in a metal when it is mined, refined, and purified. This energy varies from one metal to the next. Those metals that require the most energy to break the bonds of its oxides and silicates—for example, magnesium and aluminum—have a high store of energy, whereas metals such as gold, silver, and copper have a lower energy. The energy needed to create the metal from the minerals is stored, in a sense, and released as the metal corrodes and becomes a mineral form again. Those metals with higher energy give in quickly to the forces of nature that want to return it to an ore.

TABLE 7.1 The electro-potential of intermetallic compounds (3% NaCl sol.).

Intermetallic compound	Electrical potential
Al_3O_2	−0.51
$CuAl_2$	−0.372
$FeAl_3$	−0.14
$MnAl_3$	−0.18
$NiAl_3$	−0.468
Mg_2Si	−1.268

Aluminum, however, may be a metal of high energy but when it forms an oxide on its surface, it becomes very stable and resists changes in most environments where art and architectural surfaces are found. Aluminum seeks oxygen so readily because of the three electrons in its outer shell. Exposure of aluminum to the air causes a near instantaneous oxide. So rapid is the oxidation that it is safe to say you never see aluminum that has no oxide on its surface. If there is oxygen around, it grabs it and bonds with it.

$$Al \rightarrow Al^{3+} + 3e^-$$

When aluminum is exposed to the atmosphere, an oxide immediately forms on the surface. This oxide is approximately 4–10 μm thickness and is composed of a mineral form of gibbsite. Gibbsite is the natural mineral form of aluminum hydroxide. It has three polymorph forms, one of which is known as bayerite. Bayerite will develop as water is absorbed into the amorphous layer of aluminum oxide to form hydroxide on the surface.

$$Al \rightarrow Al^{3+} + 3e^- \qquad \text{Reduction } \tfrac{3}{4}O_2 + 1\tfrac{1}{2}H_2O \rightarrow Al(OH)_3 \text{ (Gibbsite)}$$

The color of bayerite is white to white-gray, depending on the amount of other contaminants that are absorbed into the amorphous layer.

This oxide film on aluminum provides an enormous protective barrier in nearly all environments. The major exception is the alkali environment, where aluminum forms soluble aluminates that sluff off the surface. In these conditions, aluminum will rapidly corrode.

NATURAL WEATHERING: INFLUENCES ON PERFORMANCE

The initial exposure of aluminum, regardless of alloy, will form a thin oxide film on the surface the second it is exposed. This film continues to grow and, after a few days of exposure, the growth slows way down. The thin oxide film is invisible and unreactive to the common components present in the atmosphere.

The aluminum surface exposed to the environment undergoes hydroxylation of the oxide film as moisture is absorbed from the air. Most metal oxides are hydroxylated at room temperature when moisture is present. This is a slow reaction that occurs from moisture and moisture vapor from the air collecting on the surface and reacting with the aluminum oxide.

Depending on the alloy, natural development of the oxide goes from light to dark over many years of exposure. Natural aluminum begins with a bright, white appearance. As it weathers and absorbs oxygen and water vapor from the air, it darkens to a light gray color. Zones of the surface, edges, and corners that dry slower will darken further, often developing a thick, grayish substance along exposed edges as a hydroxide of aluminum develops. The surface will get slightly coarse, possibly exhibiting some pitting, but the rate of change will start to diminish after the first few years of exposure. Aluminum will undergo the most severe corrosive attack, regardless of the environment, in the first one to two years; after that, the rate of corrosion slows considerably. Marine, urban, or

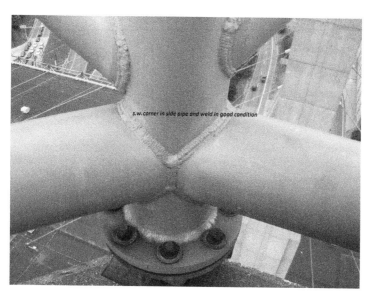

FIGURE 7.2 Image of mill finish aluminum; A96063-T6 tube exposed for 20 years.

rural environment exposures will have their most significance effects in the beginning exposures. In marine environments, aluminum could experience pitting to a depth of 150 µm in the first few years of exposure, yet only see 200 µm total after 20 years of exposure (Figure 7.2).

If there are areas where water is allowed to collect or where moisture becomes trapped between surfaces of aluminum, the oxide will thicken and darken. Additionally, if the aluminum has been abraded and the oxide removed and allowed to get wet, it will darken. This dark stain cannot be easily removed. The reason it is dark is because of the fine, powdery form of the oxide and the way light does not reflect well from the thickened rough surface.

Installed correctly (free of crevices and recesses where water will collect), aluminum will perform very well in most environments. After five to seven years of exposure the corrosion rate drops significantly as the oxy-hydroxide forms on the aluminum surface. The surface will darken and lose some of the reflective luster but still maintain a strong metallic appearance.

When aluminum is exposed to oxygen or oxidizing agents, a very thin, strongly adherent film develops. This film changes the corrosion behavior in a fundamental way and dramatically lowers the chemical activity of aluminum.

The initial exposure of aluminum, regardless of alloy, will form a thin oxide film on the surface the second it is exposed. This film will continue to grow, after a few days of exposure, slows way down.

MARINE ENVIRONMENT

Aluminum used in art and architecture is always an alloy, made up of various added elements to achieve mechanical characteristics pure aluminum is unable to achieve. These alloying elements, albeit in very tiny amounts, can influence the corrosion resistance performance of the aluminum. Depending on the process used to finish the surface of the aluminum, there will be small traces of these elements on the surface. For, instance, magnesium is a common element added to the aluminum alloys used in art and architecture, the A95xxx alloys. Whether the aluminum is in a wrought form or cast form, there will be trace amounts of magnesium-aluminum oxides on the surface. Localized electrical currents will develop between differences in electrical potential of these different compounds and the surrounding aluminum when the surface is exposed to an electrolyte. Rain is an electrolyte, so we can always expect the development of tiny local differences in electrical potential. In the case of magnesium, it would usually develop into the anode while the aluminum in the local region would be cathodic, which in turn would make the magnesium sacrificial to the aluminum.

Not so with some of the other alloys, which have elements such as copper and iron. These alloys if exposed on the surface could develop into small localized cells where the aluminum is the anode in this event. It is these localized regions of polarity that develop into all forms of corrosion that occur on aluminum alloys.

Saltwater has a pH of 8.1, which makes it a mild alkali. Aluminum will generally perform well in exposures where the pH is from 4.5 to 8. Seawater is just outside of this range. Salts will deposit on the surface of aluminum in marine environments and can concentrate into powerful electrolytes when condensation forms. If the conditions are right, this will lead to localized pitting of the aluminum surface. The appearance of the aluminum will change, beginning with a darkening of the surface, followed soon by a fuzzy amorphous growth where pits occur. Anodizing helps slow the process, but surface appearance changes can be expected (Figure 7.3).

The biggest drawback to aluminum in salt environments is from galvanic corrosion, also known as bimetallic or dissimilar metal corrosion. Aluminum will act as the anode in exposures where other metals are in close proximity and where there is an electrolyte and oxygen present. The chloride ions present in seawater and saltwater make a very good electrolyte. Aluminum will corrode as it sacrifices itself to most other metals. Zinc and magnesium are the exceptions, and they would sacrifice to the aluminum in marine environments. That is why zinc anodes are used on the underside of boats to act as sacrificial metals for the other metals, including aluminum.

If there are iron or copper particles or ions present, severe corrosion can be expected on the aluminum. The iron and copper particles will initiate corrosion cells on the surface of the aluminum. Pitting will develop around the corrosion cells and this may lead to eventual perforation.

Exposures of certain alloys of aluminum, particularly the wrought forms of extruded aluminum, can undergo a very destructive form of intergranular corrosion known as exfoliation. Initially there are visible pits on the aluminum surface that form into a more compact grouping that appears to be elongated in the direction of the grain or length of the extrusion. As the corrosion advances along the length of the extrusion, it separates the aluminum into layers, like rotting wood or the wet pages

FIGURE 7.3 Anodized bridge handrail exposed to the environment for 18 years; changes occurring at different rates.

of a book. Once this occurs, it is irreversible, and the damaged aluminum needs to be removed. It will tend to feed on itself as moisture is absorbed and the aluminum hydroxide expands. To aid in preventing this from occurring, the correct heat treatment of the aluminum alloys should be used, and this should be followed by anodizing to add a thick, inert oxide layer on the surface. The ends of the aluminum parts should be anodized as well. It is important that the finished part be anodized or painted after cutting to length in order to provide a protective coating on the ends of the extrusion.

URBAN ENVIRONMENT IN THE SOUTH

For the most part, aluminum can be expected to perform well in urban environments in the South. An occasional cleaning, once or twice a year, is enough to keep aluminum surfaces performing and appearing in good shape for years. Perhaps the most damaging exposure can come from window- or stone-cleaning substances allowed to remain on the aluminum surface. These are often alkali or acidic fluids that can damage the aluminum. Rinsing them off immediately will limit any potential damages. Aluminum is especially susceptible to alkalis; even dilute concentrations can corrode aluminum.

Aluminum exposed to urban environments develops a hydrated aluminum oxide over the surface. The reflective appearance of the metal will dull as the oxide develops and uniform corrosion occurs across the surface. Urban environments can often have a lower pH—that is, more acidic—due to the pollutants in the air from a higher density of combustion particles in the air. Combustion products contain carbon monoxide, sulfur dioxide, nitrogen oxide, and carbon dioxide. Nitrogen oxide and sulfur will mix with the surface moisture on aluminum and create a stronger, acidic, electrolyte solution. Sulfate ions will react with the hydrated aluminum oxide and form aluminum sulfates, which are soluble and will wash away as rain and moisture collects. Eventually they will form insoluble sulfated hydroxides that are darker and thicker. The corrosion product that forms is an amorphous clump that adheres to the surface. Usually a pit has formed below this corrosion product.

The corrosion is usually spotty. It can be removed but this will leave a shiny spot on a uniformly dull surface. The rate of corrosion should slow as the aluminum darkens over time. The process moves quickly at first but as exposure continues and the oxide thickens, occasional spots will develop and small lumps of aluminum hydroxide sulfates will form on a few edges.

Allowing moisture to sit on the surface will create a dark spot of concentrated hydroxide, which will be difficult to impossible to clean. Careful design and a cleaning regimen will diminish this condition.

URBAN ENVIRONMENT IN THE NORTH

The urban environment in northern cities, particularly in the United States, experiences high exposures to anti-icing salts. Anti-icing salts contain chlorides and their ability to lower the relative humidity generates a corrosive attack on most metals. Aluminum exposed to urban environments has the double-edged attack of the chloride salt from anti-icing practices and the pollutants from combustion processes in urban environments, as well as airborne iron particles from construction in nearby areas.

The chloride will create dark gray to black spots as pits develop in the aluminum. Continued exposure to corrosive substances without any thorough cleaning will etch into the aluminum and can cause more severe intergranular effects along edges and bases of surfaces where moisture remains longer. The surfaces can be cleaned but the pits will remain. If intergranular corrosion occurs, the metal will be rendered useless as it frays and dissolves away.

Alloys of aluminum containing copper and manganese should be avoided unless they are coated with other, less reactive alloys in a process known as alclad. They can also be painted with a sound enamel and a chromate pretreatment. The A95xxx series alloys have shown better results in these exposures. Anodizing helps slow the process, but if chloride salts from anti-icing processes are allowed to remain on the surface, you can expect these alloys be attacked as well. For anodized surfaces concentrated salts can damage the seal and eventually the oxide. For clear anodized aluminum

FIGURE 7.4 Corrosion on protected underside in a semiurban environment; 12 years exposure.

this can amount to white spots while for color anodized, the salt can interact with the metallic salts in the pores and alter the color. The solution is to engage in a maintenance program where the surfaces are washed a few times a year, particularly in the springtime, to remove any anti-icing salts from the surface before they have an opportunity to attack the metal. If possible, wash the surface with deionized water. Deionized water removes the salts more effectively than tap water and reduces spotting (Figure 7.4).

RURAL ENVIRONMENT

Aluminum will perform very well in the rural environment where the occasional rain will remove general dirt and airborne grime. An infrequent rinse with clean water will further aid in keeping aluminum surfaces looking and performing well. Whether natural, bright finished aluminum or the matte appearance of anodized aluminum, surfaces constructed of this metal in the rural environment can be expected to provide decades of service. The alloy is less critical in rural environments. One simply needs to exercise an occasional cleaning to keep the aluminum surface looking lustrous.

The more polished surfaces perform the best. The smooth surfaces shed dirt and other pollutants that may alight on the metal. The coarse surfaces hold dirt and airborne deposits. Cleaning will be required if expectations are for the luster of aluminum to continue.

CORROSION OF ALUMINUM SURFACES

The corrosion resistance of aluminum is excellent in most environmental exposures. Aluminum forms a thin oxide layer nearly instantaneously when exposed to air. This oxide layer is tightly bound to the base metal and affectively seals the core aluminum from any further reaction with the environment. Unlike steel, the oxide layer on aluminum forms a chemical bond that is very tightly knitted onto the surface. Steel, when it oxidizes, has a tendency to change in volume. The iron oxide disconnects from the base metal, made evident by the flaking that occurs with heavy rust development. When this occurs, more base metal is exposed to the atmosphere and the process continues. While with aluminum the oxide that rapidly forms on the surface is very adherent and, if removed, returns quickly to protect the core material. It does not sluff off the surface or stain adjacent materials.

The oxide that forms on aluminum is stable in the pH range of 4–8. That range is typical of most urban and rural environments in the world today. Aluminum can be expected to perform well in most exposures because of this oxide. However, there are instances when aluminum will experience corrosive environments that have an occasion to fall outside this range. The oxide will break down at extremes of pH. Marine environments and heavy industrial environments are two such categories of exposure that can exacerbate the corrosive attack on aluminum. The surface of the aluminum can show signs of pitting after several years of exposure to marine or chloride environments. Usually these small pits accompany a gray powdery substance. A film of water, via humidity, coats the surface of aluminum on exterior exposures. Pollutants on the surface, such as sulfur dioxide and chlorides, become ionized in the film of water. While moisture is present an electrochemical mechanism can be initiated and attack the surface of aluminum and create small pits.

Similar to other metals as they corrode, they tend to develop mineral substances on their surfaces. The mineral nacrite, $Al_2Si_2O_5(OH)_4$, is associated with the development of the oxide layer on the surface when aluminum is exposed to marine environments. This is a claylike mineral, gray in color. Other minerals that form when aluminum is exposed to corrosive attack are diaspore (AlO(OH)), as well as aluminum hydride (Table 7.2).

It has been found that when exposed to the environment aluminum corrosion decreases with time. This is due to the development of these mineral-like oxide layers on the surface. Tests performed in Panama on A91100 and A96061 alloys both corroded at decreasing rates with time. Over 16 years the tests showed the rates of corrosion amounted to 67 and 63 g m^{-2}, respectively.

TABLE 7.2 Types of corrosion.

Type of corrosion	Identifying symptom	Suggested remediation
Uniform corrosion	Gray, blotchy surface. Can be on sheltered areas.	Clean the surface and set up a regular washing protocol.
Galvanic corrosion or bimetallic corrosion	Initial darkening and fuzzy gray matter at connection. Pitting and dissolution of metal	Remove the corrosion products. Separate the metals by painting the cathodic region.
Pitting corrosion	Small dark pits randomly dispersed across a surface	Clean corrosion products from the surface and institute a cleaning regimen
Exfoliation corrosion	Flaking layers, layers like a wet book or rotting wood	Remove the damaged area. Seal the balance.
Stress corrosion cracking	Crack in the metal, usually accompanied with a dark area of oxide	Remove the damaged area. Replace the part and relieve the stress.
Corrosion fatigue	Crack in the metal across grain	Reduce or eliminate repeated stress conditions. Remove damaged area.
Crevice corrosion	Excessive pitting under seals or washers and laps.	Eliminate the crevice between the surfaces. Replace broken seals.
Filiform corrosion	Small wormlike tracks under organic and inorganic coatings-severe forms delaminate coating	Grind off the paint and oxide. Repaint with a zinc-rich primer and good overcoat. Exercise care around fasteners.
Fretting corrosion	Gouging or ripping of surface	Separate surfaces with lubricant or other material

The Pourbaix[1] diagram (Figure 7.5) indicates passivity when the pH of the environment is between 4 and approximately 8.7. Local environments at the surface may be outside of this range if salts collect around crevices or fasteners. When the environment the aluminum is exposed to falls outside of this pH range the oxide layer can break down and reformation of the protective layer can be impeded. The sloping dashed lines are the reduction-oxidation potentials of a solution. These lines indicate an electron transfer as the pH changes.

---

[1] A Pourbaix diagram is also known as a potential pH diagram. It is used to map out theoretical stable and unstable regions of an electrochemical exposure of a material by plotting pH against electrochemical oxidation reactions of an element. The diagram provides information on aluminum in three states—active, passive and immunity—or thermodynamic stability.

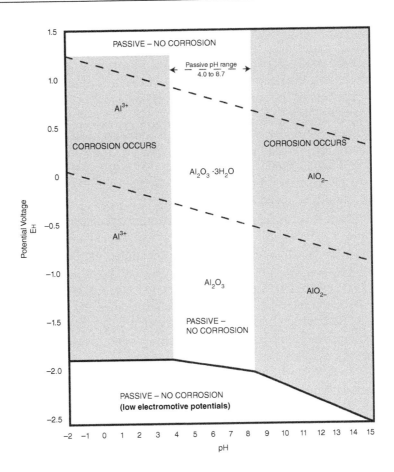

FIGURE 7.5 Pourbaix diagram for aluminum.

Uniform Corrosion

In describing uniform corrosion for aluminum, it could best be understood as general surface oxidation or tarnish. Aluminum will not tarnish as readily as most metals and, if anodized, will not tarnish at all. Aluminum will absorb water from the atmosphere, and this will thicken the outer layer. The shiny aluminum surface will over time dull slightly as the oxide grows. Handling of aluminum will leave fingerprints with oils that can actually etch the surface. The coarser the finish on the aluminum, the more difficult it can be to remove the fingerprints. Anodized aluminum surfaces will take fingerprints, but not nearly to the level of nonanodized aluminum. Cleaning the surface of anodized aluminum is significantly easier.

Galvanic Corrosion

Galvanic corrosion or bimetallic corrosion is the term used to describe the electrochemical reaction when two or more metals are in proximity to one another. The condition has also been described as "dissimilar metals" and denotes the condition when two metals are in close proximity and exposed to the environment (Figure 7.6).

Corrosion experienced when two dissimilar metals are connected is not unlike the corrosion that occurs when metals are not coupled, but the process is sped up for the more anodic metal or, in the case of the more noble metal, is slowed way down or eliminated altogether. The galvanic relationship of metals is a powerful tool when designed for and utilized correctly, or it can be a disastrous condition when overlooked.

For galvanic corrosion to occur, several key conditions must be met. Eliminate any one of these conditions and galvanic corrosion will not happen. Oxygen must be available to the cathode. This usually is supplied from the electrolyte as dissolved oxygen and forms a hydroxide. In the absence of oxygen, hydrogen ions form at the cathode. This can create a different issue with cathodic corrosion but in the context of art and architectural surfaces, oxygen is usually available.

There must be an electrolyte that provides the connection of the circuit. If the environment is dry, the electrolyte is eliminated, and galvanic corrosion will not occur. If there is a barrier, such as a coating of nonconductive material or nonconductive oil, the flow within an electrolyte is prevented (Figure 7.7).

There must be flow of electrons from one metal, the anode, to the cathode. This is where the two metals are in contact. This is dependent on the difference in electrical potential.

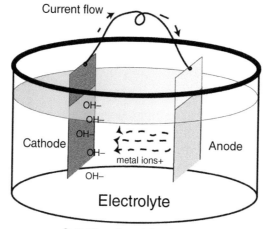

Current flows through the electrolyte from the anode to the cathode

FIGURE 7.6 Diagram of a corrosion cell set up.

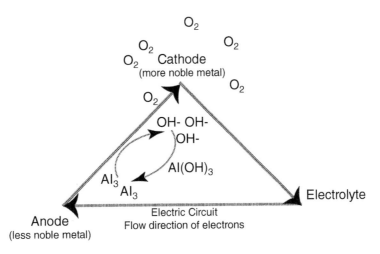

FIGURE 7.7 What is needed for galvanic corrosion.

All metals have an electrical potential. The electrical potential is the tendency of a metal to give up electrons when submerged in an electrolyte. Table 7.3 is indicative of the polarity of metals in flowing seawater. The electrical potential is a measure of a particular metal submerged in an electrolyte and compared to the electrical potential of a known electrode.

This chart is also referred to as the galvanic series. It is derived by measuring the electrical energy released when it is immersed in a specific environment, in this case seawater. The relationship on this chart and the electrical potential measured can change, depending on the environment tested. If different combinations of saltwater concentration, temperature, moving or stagnant water, or a different solution altogether, the voltage readings will be different. There are many factors that affect the electrical potential of a metal.

The arrangement of metals in the chart should be considered as the relative tendency of one metal corroding sacrificially when in contact with the other metal lower down in the chart. The metals lower down in the chart, the less negatively charged metals, are considered more "noble" than the metals above them. A voltage difference is necessary for the flow of electrons to occur. The larger the difference, the greater the potential and the more rapid the corrosion of the less noble metal. It is important to note, the galvanic series and the differences in potential do not tell the entire story on galvanic corrosion. The magnitude of the current flowing between the two metals is more critical than the position on the chart.

Aluminum has a relatively high negative potential. This means it has the potential to be consumed (corroded) when in the presence of most other metals. Zinc and magnesium are two metals that are higher in the electrical potential than aluminum; they would be corroded by aluminum if the right circumstances were met. When aluminum forms the thick aluminum oxide or when the aluminum is anodized, the Al_2O_3 layer makes the surface of aluminum more electrically positive and actually would move aluminum down the chart.

TABLE 7.3 Electromotive scale.

Anodic polarity	Electrical potential of various metals in flowing seawater	
	Voltage range	
The more active end of the scale—least noble metals	−1.06 to −1.67	Magnesium
	−1.00 to −1.07	Zinc
	−0.76 to −0.99	Aluminum alloys
	−0.58 to −0.71	Steel, iron, cast iron
	−0.35 to −0.57	S30400 stainless steels (active)
	−0.31 to −0.42	Aluminum bronze
	−0.31 to −0.41	Copper, brass
	−0.31 to −0.34	Tin
	−0.29 to −0.37	50/50 lead-tin solder
	−0.24 to −0.31	Nickel silver
	−0.17 to −0.27	Lead
	−0.09 to −0.15	Silver
	−0.05 to −0.13	S30400 stainless steels (passive)
	0.00 to −0.10	S31600 stainless steels (passive)
	0.04 to −0.12	Titanium
	0.20 to 0.07	Platinum
The more noble end of the scale	0.20 to 0.07	Gold
	0.36 to 0.19	Graphite, carbon

CATHODIC POLARITY

Corrosion Parameters

For aluminum, dissimilar metals occur most often with fasteners, clips, and steel or stainless steel support girts. They can also occur when aluminum framing is inserted in a copper or steel clad structure. For aluminum, dissimilarity can also occur when different alloys of aluminum are used or when different heat treatment processes are used on the same alloy; however, this will require other conditions to be met before anything potentially severe will occur. For instance, aluminum alloys that contain copper are more noble than pure aluminum. Thus Alclad, the coating of one aluminum with another, often involves more pure alloys as the coatings of less pure alloys. The pure alloy has good corrosion inhibition in many atmospheric exposures, but as a coating of less pure alloys, the pure coating also acts sacrificially and is more anodic than the more alloyed forms.

Corrosion always occurs at the anode as hydrogen gas is released from the electrolyte around the anode. Hydroxyl ions form at the cathode and these migrate to the anode and react with the dissolved metal at the anode surface and form metal hydroxides or hydrated metal oxides. It the oxides are insoluble, they deposit on the anode surface. This slows down the rate of corrosion. This happens when aluminum is anodized. Anodizing is a form of controlled corrosion of the aluminum surface.

Several conditions play a role in whether galvanic corrosion will occur with aluminum and another metal. If the conditions are met, then the question is how severe the corrosion ultimately will be.

Parameters that will have an effect on corrosion

- Electrical potential differences between the metals
- Electrical connection of the two metals
- Ratio of areas—cathode to anode
- Presence of an electrolyte that bridges across the two metals
- Conductivity of electrolyte
- Oxygen

There needs to be a differential in electrical potential for a current to flow between two metals. The current flows from the metal that is more negative or anodic to the metal that is more positive. The anode is consumed or erodes as it dissolves into the electrolyte. The greater the electrical potential, the more significant the current can be in the right electrolyte and the faster the effect of corrosion on the more negative or anodic metal. Thus, the galvanic series chart is helpful in determining the order and magnitude one might expect in the coupling of metals. But there are a number of other conditions that will have an influence on the rate or corrosion of two metals in proximity.

The metals must be connected in some fashion to one another. They can be in direct contact or, as discussed later in this section, one metal can be connected by ions in solution flowing onto a less noble metal. Another way might be via a grounding line connecting the two dissimilar metals through a second material. The metals might be insulated from one another by a nonconductive gasket, but connected elsewhere by a grounding connection.

Three main conditions must be met or galvanic corrosion will occur. These measures are (1) eliminating oxygen from reaching the cathode, (2) insulating between the cathode and anode to prevent electron flow, and (3) eliminating the electrolyte. Without these, the circuit cannot occur (Figure 7.8).

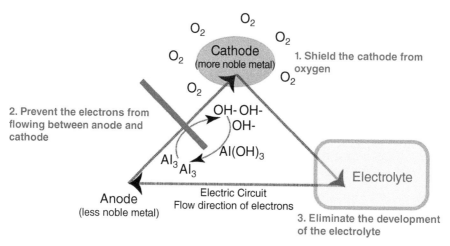

FIGURE 7.8 How to prevent galvanic corrosion.

Oxygen

Oxygen is necessary for aluminum to corrode. A condition that lacks oxygen will not corrode in most all art and architectural exposures of aluminum. Aeration and flow rate of the electrolyte play an important role in the rate of corrosion due to bimetallic coupling. If oxygen diffusion to the aluminum surface does not occur, corrosion will slow. Aluminum surfaces will develop a stable oxide film in flowing water and this oxide will thicken with time and become more stable. As the waters flow over the surface, oxygen diffuses out of the electrolyte onto the aluminum surface. The rate of diffusing will decrease over time as the oxide forms; this will slow the rate of corrosion.

Aluminum alloys that contain magnesium (A95xxx alloys) form a protective oxide film in most clean seawaters. If the area ratio is sufficient, these magnesium-bearing aluminum alloys have shown success when coupled with steel in marine environments.

Ratio of Areas

Galvanic corrosion is dependent on the relationship of area of contact or, more specifically, the ratio of area of cathode to area of anode. The larger the cathode, the greater the galvanic current, and the smaller the area of cathode, the less current will flow. For example, if a stainless steel fastener is used to fix a large aluminum cross-section or panel to an aluminum frame, the area of stainless steel in relationship to the area of aluminum is very small. The galvanic current will be constant between the two metals, but the corrosion per unit of area of the aluminum is low because the current density is low. The stainless steel fastener is usually desired for the shear strength it will provide. Stainless steel is more noble than aluminum and the area of aluminum is significant in relationship to the small stainless steel fastener. The stainless steel will be protected by the aluminum anode and corrosion of the fastener, even when exposed to a strong electrolyte, will be negligible. At the same time, the aluminum will experience very little corrosion because of vast differences in mass. With the opposite condition, where the mass of aluminum is small compared to the mass of a more noble metal, intensive and rapid corrosion can be experienced.

This is one of Faraday's laws, where a given current passes between the anode and cathode in a galvanic cell at a proportional rate. If, for example, the cathode area is 10 times larger than the anodic metal area, then the current will be 10 times as great passing through the anodic metal, and corrosion of the anode will be rapid (Figure 7.9).

Rule 1: Ratio of cathode/anode area should be a small value

Area ratio plays a role in painted and coated aluminum surfaces. If there is porosity in the coating, or if the coating has been damaged, abraded, or pierced, the exposed aluminum can be a very small anodic zone and the ratio of cathode/anode can increase, causing a more rapid corrosive attack on the aluminum at the breach in the coating.

Therefore, it is better to coat the cathodic metal rather than the aluminum. This keeps the ratio of areas down in the event of a breach in the coating.

Rule 2: Coat the cathode

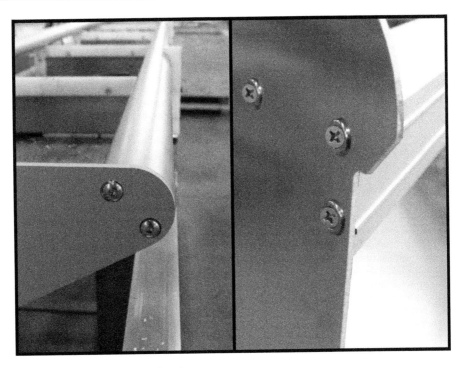

FIGURE 7.9 Stainless steel fastener in aluminum.

Electrolyte Strength

An electrolyte is an electrically conductive solution. Pure water is a poor electrolyte, but if pollutants, such as chlorides or sulfides, enter the solution, they can make powerful electrolytes. Acidic solutions or alkali solutions can increase the conductivity of the electrolyte. Seawater or dissolved anti-icing salts are very strong conductors of electrical current, and there can be an increased rate of corrosion on the area where the anodic metal meets the more noble metal.

A solution increases its ability to conduct electricity by increasing the ionization of substances dissolved in them. An electrolyte develops as the dissolved compounds establish positive and negative ions in the solution. The solution becomes capable of conducting electricity by the migration of positive and negative ions within an electric field. The strength of the electrolyte as measured by its ability to conduct electricity is dependent on the density of ions per unit volume and what is referred to as drift velocity of the ions. The drift velocity is determined by the size of the ion and the strength of the electric field that develops.

For aluminum, pH levels between 4 and 8.5 are considered safe in most circumstances. However, in the presence of more noble metals, a pH of 4–5, an increase in the corrosion rate of the aluminum will occur, particularly if the more noble metal is copper.

Thin films of moisture, such as dew or condensation, carry a lower electrical conductivity, but they can cause a significant level of corrosion on the areas where the two dissimilar metals meet if the difference of the electrical potential is significant. This increases the electrical field and thus the drift velocity of the ions (Table 7.4).

Aluminum can be corroded when moisture passes over or through a metal, such as copper piping or a copper guttering. Some of the copper ions will pass into solution as fresh waters travel over the surface. If this water is passed over aluminum or held in an aluminum gutter or tank, the copper ions will redeposit onto the aluminum. Subsequent corrosion will occur to the aluminum as small pits will develop. Copper condensation piping on a rooftop draining over an aluminum surface will deposit copper particles and etch the aluminum. This is not the case with stainless steel surfaces draining onto or over aluminum surfaces, but steel corrosion particles can stain the aluminum and eventually develop into corrosion cells on the aluminum surface.

TABLE 7.4 Electrical conductivities of various substances.

Electrolyte	Electrical conductivity	Relative conductivity
Pure water	0.0002 mho m^{-1}	Poor
Alcohol	0.0003 mho m^{-1}	Poor
Salt solution (NaCl)	20 mho m^{-1}	Good
Oil	1×10^{-14} mho m^{-1}	Poor (good insulator)
Aluminum	3.8×10^7 mho m^{-1}	Excellent
Copper	1×10^8 mho m^{-1}	Excellent

> Rule 3: Avoid exposing aluminum to strong electrolytes.

> Rule 4: Avoid water passing from more noble metal to aluminum surfaces.

Pitting Corrosion

In marine environments, the alloy type of aluminum is important for corrosive resistance performance. Marine environments are rich in chloride salts. The chlorine ion will set up very strong electrolytes if moisture is present. Marine exposures typically have a higher relative humidity and, when coupled with the chlorine ion, the surface more rapidly develops a uniform layer of hydrated aluminum oxide that dulls the surface reflectivity. The hydrated oxide helps protect the aluminum but the deposits of salts, if allowed to remain on the surface develop into aluminum chloride hydroxides. This has a darker appearance with small spots forming where chloride concentrations are present. Pitting of the surface usually follows as small corrosion cells develop. The small spots become fuzzy, grayish white spots of amorphous aluminum compounds. These fuzzy compounds are no longer part of the metal body but become porous substances that can retain moisture and other pollutants, compounding the problem. Pitting from anti-icing salts has a similar effect. As the dusty white corrosion particles are removed, pits in the aluminum will become apparent (Figure 7.10).

Pitting resistance of various aluminum alloys from most to least.

A95052	*Most resistance*
A96063	
A91100	
A93003	
A94043	
A96351	*Least resistance*

Intergranular Attack

Saltwater environments are near neutral in pH, and as such they would not corrode aluminum. The marine environment though, will provide the necessary ingredients for corrosion to occur. For aluminum, as with other metals, the alloy selection for these environments is paramount.

Aluminum alloys that contain copper should not be considered in marine environments. This would be the A92xxx series alloys unless they are clad with more pure aluminum, such as in the case of alclad sheets or plates, or if they are coated with a zinc rich paint.

There are several alloys that have been considered as marine grade alloys due to their alloying makeup.

FIGURE 7.10 Aluminum in a marine environment. Areas of corrosion developing. Fuzzy black pits developing, strongest around welds.

Marine Grade Alloys of Aluminum
A96061
A96063
A95083
A95086

These alloys tend to hold up better in most marine environments. Additional protection, however, should be considered when using aluminum in high-chloride environments. Note, the effects of salt deposits from coastal exposures drop off rapidly as the distance from the coast increases. Within 1600 m of the coast, further protection of the aluminum is advised (Figure 7.11).

Anodizing, painting, and, in special cases, cathodic protection may be warranted. Cathodic protection involves connecting the aluminum to the negative pole of a DC voltage source. This keeps the electromotive potential low but adds maintenance. For building structures, this is rarely utilized. For various art installations, it could be warranted. Certain intergranular corrosive conditions that can develop are irreversible, such as exfoliation. These will be discussed in further detail in the section titled "Exfoliation," further on in this chapter.

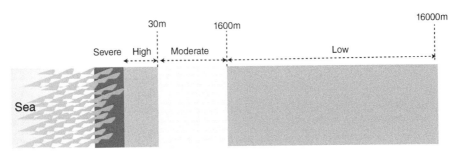

FIGURE 7.11 Distance from sea coast and salt exposure.

When exposing aluminum alloys to the marine environment, even the marine grade variety of alloys, the designer should recognize galvanic corrosion possibilities. Aluminum will sacrifice itself galvanically to most other metals and with the strong electrolyte developed by seawater exposure, this can become critical as corrosion cells develop.

Intergranular corrosion is the main form of attack on an aluminum surface in marine and salt intensive environments. Pronounced use of anti-icing salt in northern climates can create a significant problem to aluminum similar to that of intense marine environments. Intergranular corrosion is characterized by the occurrence of numerous small grayish spots. These spots are the corrosion product generated at the intergranular boundaries of the metal grains. Localized polarity develops where some areas on the surface are more noble than other areas (Figure 7.12).

Simply put, the minute areas on the surface of the metal have a different electrical charge. This leads to the development of a galvanic cell when an electrolyte is present.

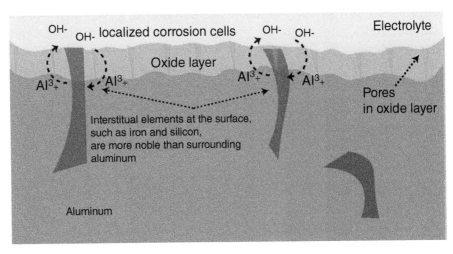

FIGURE 7.12 Diagram of cross-section, intergranular attack.

These darkened spots are amorphous. They lack any real structure and grow out from the surface. They are representations of the minerals bayerite and gibbsite. They are composed of hydroxides of aluminum, $Al(OH)_3$. When the deposits are removed, a small pit is in the surface.

The darkened spots usually manifest across the surface of the aluminum in an irregular pattern. They are always the result of exposure of the aluminum surface to a corrosive salt, such as chlorides or sulfides. They also can form if the water is highly alkaline, when the pH is greater than 8.7.

These spots of graying corrosion product can be cleaned from the surface. The pits will remain, but the discoloration can be removed as long as the corrosion has not become too severe. This type of corrosion can spread across a surface into a more uniform attack and actually turn the aluminum to a powdery dust creating holes and ragged edges in extreme cases.

Exfoliation

When intergranular corrosion is severe, a process called exfoliation occurs. Exfoliation corrosion is the name given to severe corrosive attack on aluminum surfaces. Exfoliation is more common in the aluminum alloys containing copper and the alloys containing copper-magnesium-zinc as alloying elements.

Exfoliation is a damaging and irreversible corrosion condition that can develop around edges and fastener holes in aluminum. Exfoliation is extremely destructive to aluminum of all forms, extrusions, plates, tubing, and even sheet. This type of corrosion attacks the grain boundaries of aluminum and causes a structural splitting that appears like the flayed pages of a book. The corrosion products selectively attack weaker regions of the metal and expand, while other portions remain relatively corrosion free. The corroded regions become weak and expand as they turn to powder. Exfoliation begins as the grains are etched out. Aluminum alloys in the wrought forms have their grains arranged in stratified layers. These layers influence the strength and mechanical behavior, but they can also be subject to differential electrochemical regions where minute cathode and anode relationships develop. As this occurs, the aluminum grains hydrate inward and form hydrated aluminum oxides between grains. Hydrated aluminum oxide occupies a greater volume and thus expands making a flayed appearance, as shown in Figure 7.13.

When exfoliation occurs, it cannot be repaired. The aluminum part must be cut out and removed. Certain aluminum alloys perform better in environments where intergranular corrosion can develop. These environments are high-humidity marine environments and where anti-icing salts are used. Using the A95xxx series alloys is recommended. These have better intergranular corrosion resistance.

- A95083 H116, H321, and H323
- A95086 H32, H321, and 0 temper
- A95454
- A95052
- A95456

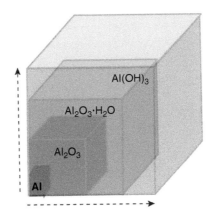

FIGURE 7.13 Change in volume as the aluminum oxide hydrates and expands.

Those aluminum alloys that contain copper and manganese should not be used in marine or anti-icing environments unless they are adequately protected by coatings. When exfoliation is a concern, the edges of the parts must also be coated.

Filiform Corrosion

Filiform corrosion occurs just beneath organic or metallic coatings on aluminum as well as other metals. Filiform corrosion can begin at small defects in the coating where moisture and oxygen can reach the metal. Lacquers and field applied paints are most susceptible because there is always some porosity in the coatings, due to oils or minute pores in the coating layer. Filiform appears as random threads that can resemble worm tracks on the surface of wood. The track starts at the pinhole in the coating or at an edge where a fastener pierces the coating. It can be a series of tracking lines moving outward from the hole or appear as a blister where the paint looks as if it is bubbling up. Corrosion begins as the aluminum forms a thick oxide that migrates under the coating in a shallow channel that forms as the metal oxidizes. As it progresses over time, large delaminated sections appear as the paint or coating separates from the surface. As the paint is raised, a powdery oxide on the metal surface remains (Figure 7.14).

Filiform corrosion occurs in high-humidity environments, both interior and exterior, where a porous coating has been applied to the aluminum. At first it seems minor and not especially damaging. Sealing over the surface may stop it from growing but the powdery residue below the filiform track will remain and eventually cause the paint to flake.

As filiform corrosion develops, the head of the track becomes the anode in the relatively confined space, usually smaller than a millimeter. At the head of the track, the pH is lower and there is a lack of oxygen, while behind the head the region becomes cathodic with more oxygen available from the pore (Figure 7.15).

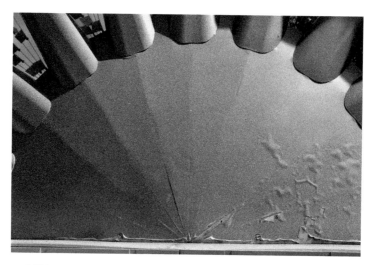

FIGURE 7.14 Image of filiform corrosion. Occurs over time and grows under the paint coating.

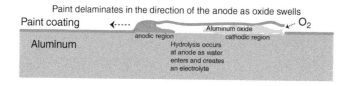

FIGURE 7.15 Cross-section of filiform corrosion under a paint coating.

It can be difficult to locate porosity in a coating on aluminum. Most oven-baked factory-applied coatings are sound. They have pretreatments that ensure a good bond. Older coatings that were applied in the field or using paints that were not fully cured at the time of application can be vulnerable to filiform corrosion.

Stress Corrosion Cracking

Stress corrosion cracking is a type of intergranular corrosion that can occur when an aluminum part is subjected to stress induced from tensional forces while exposed to a corrosive environment. It is an interaction of mechanical loading conditions and environmental factors. Stress corrosion cracking in aluminum can be both along and across grain boundaries.

High-purity alloys of aluminum are not susceptible to stress corrosion cracking.

Most art and architectural uses of aluminum are engineered to keep the stress conditions low. Most designers consider exposures of a magnitude within the service mechanical range of the alloy used. However, when exposed to corroding environments, the necessary stress conditions to cause

FIGURE 7.16 Damage from stress and internal corrosion.

a crack can be significantly lower than the design yield of the metal. Additionally, if the metal form was already under stress from forming or shaping operations, it can be more susceptible to stress corrosion cracking (Figure 7.16).

For stress corrosion to be a concern, the following factors need to be met:

- The alloy and temper must be susceptible to cracking.
    Alloys that show good resistance are A95xxx in the H116 and H117 tempers. Also, alloys of A96xxx and A91xxx have good resistance to stress corrosion cracking behavior.
- The environment where the aluminum is exposed must be corrosive.
- The stress imparted to the aluminum must be sufficiently high to create the crack.

Fretting Corrosion

Fretting corrosion occurs when two surfaces are in contact with one another. On a microscale, a surface of metal will have very small undulations called asperities. Aluminum surfaces have what is called a high stacking fault energy due to the crystal makeup of the surface. Material is transferred to the adjacent surface as it slides over the surface. This behavior is why aluminum is more subject to galling.

Fretting corrosion occurs when the two surfaces are vibrated against each other. Mechanical stress can develop and pull sections off of one and onto another surface. These, in severe instances, can rip and friction weld the metal onto the other surface.

Initially the surface may appear dark in color, but as it continues it will damage the surface.

For aluminum that is anodized, slipping is facilitated and galling is reduced. You can also eliminate the vibration and slip completely in the design. Lubrication and gasketing with a softer material can also prevent fretting corrosion.

Crevice Corrosion

Crevice corrosion is the term given to corrosion that occurs in or under concealed regions. These regions have limited oxygen or may collect pollutants due to lack of aeration and natural washing. This type of corrosion is insidious because it is not readily visible by inspection.

What happens is that a differential in electrical potential develops between the concealed, oxygen-deprived region and the general area. This causes current flow, and pitting can be initiated in the concealed region. Crevice corrosion can occur under areas where one metal surface laps over another. It can occur where a gasket seal is incomplete and corrosion particles can enter or where moisture becomes trapped.

To prevent or inhibit this type of corrosion, make sure the joints are well drained and sealed. Open them slightly to allow airflow and natural rains access to the surface or seal it entirely. Clean back any areas that appear decayed or damaged. Paint concealed areas with a zinc-rich primer. For crevice corrosion to occur, the gap must be less than 1.5 mm. Water will enter the gap by means of capillary action, and as other pollutants are brought in, crevice corrosion may begin.

Crevice corrosion is one of the more insidious forms of corrosion. A surface can look as if it is performing well from the outside only to be corroding internally in the hidden layers. Look for signs of darkening and grayish corrosion particles along the edge of the opening. These are signs that something is happening within the gap.

WATER STAINING

If water is allowed to remain on the surface of aluminum, it will generate dark stains. The depth of color depends on the thickness of the hydroxide that develops on the surface. Water stains are created by hydro-oxidation of the aluminum. Aluminum exposed at length to freshwater forms a dual layer that affects the appearance. The outer layer is a roughened layer of noncrystalline mineral similar to bachnite, AlOOH, and the inner layer is a thickened, amorphous form of aluminum oxide, Al_2O_3.

Coils and sheets of aluminum allowed to get wet will develop very adherent stains. Water allowed in between two surfaces of aluminum will stain quickly. Protective wrappings on aluminum during storage will induce stains on the surface if they trap moisture. The stains can range from white, powdery oxides or dark gray to black. Removing them is extremely difficult.

Cleaning and etching processes used to prepare metal for anodizing do not completely remove the stains. They can instead fix the stain into the aluminum surface because the stain will remain in the anodized layer. Also, much depends on the base texture applied to the aluminum. The finish of aluminum provided from the mill, such as specular and as-fabricated, cannot be restored. You would need to either completely abrade the surface to remove the oxide, then apply a finish, or chemically etch the surface to remove the stain, then follow up with a secondary finish, such as satin or glass bead.

Sodium hydroxide will etch the surface and, if the darkening is minimal, can remove most of it. Abrading the surface will remove the oxide, but selective abrading the surface—that is, not completely abrading the entire surface—can leave an appearance that is worse than the stain. Satin finishes, those with a linear or circular finish, can be chemically treated then refinished. Chemically removing the surface has shown success. Because the oxide is so unyielding, powerful acids or bases are needed, coupled with the assistance of a mild abrasive pad.

Some cleaners using dilute hydrofluoric acid can remove the stain, but they etch the surface when doing so. These treatments require safety protection for the personnel applying them as well as for the environment. Some of these cleaners are available in proprietary mixes with automotive parts stores. They are formulated to clean aluminum wheels on cars.

FINGERPRINTING

Aluminum will take fingerprints. The coarser the surface finish is, the more readily it will fingerprint. The acids in perspiration will etch into the surface and make cleaning very difficult. Anodized aluminum will fingerprint but the surface etching will be thwarted by thicker, sealed surface. Removing the fingerprint requires displacement and lifting the oils from the surface then wiping them away. One way to remove fingerprints is to use isopropyl alcohol to displace the mark, then a mild soap to capture the oils, followed with a wipe down with deionized water. If the surface has been etched from the fingerprints, a mild abrasive may be necessary. Many aluminum polishes contain a mild abrasive that can be used to bring the surface back to new.

EXPECTATIONS OF VARIOUS ENVIRONMENTAL EXPOSURES

Fresh Water

Distilled and deionized water will not have an effect on aluminum. Fresh waters are so varied in composition it is difficult to precisely determine how the aluminum will perform, but one could expect aluminum to perform well in clean fresh water. There may be localized pitting if the water is stagnant and the aluminum remains in the water. Small boats and canoes used in fresh water are often made of aluminum and hold up well without any special care or cleaning. Steam from clean, fresh water will not corrode aluminum.

If, however, iron or copper ions are present in the water, pitting will be initiated. Chlorinated water will also add to the pitting of aluminum. Anodizing and painting the aluminum surface will aid in the resistance to pitting.

Seawater

Aluminum alloys that do not have copper as an alloying constituent will perform well in seawater exposures. The pH of seawater is neutral. You can expect some localized pitting of the surface. Aluminum should be expected to perform adequately in clean seawater exposures. Salt solutions where the pH is in the range of 5–8.5 will not affect the aluminum surface. Anti-icing salts are hygroscopic and absorb moisture, which can form alkali solutions.

Polluted Waters

Waters with traces of copper ions will attack aluminum. Copper piping dripping into an aluminum vessel will corrode the aluminum. Copper ions coming out of condensation units will corrode the aluminum surface, first creating a dark streak that is difficult to remove, and eventually etching the surface. Even anodized aluminum is susceptible to copper ions. Aluminum curtainwalls below or adjacent to a copper surface must be well drained. Consider paint on the aluminum to protect it from the etching effect of the copper runoff.

Mercury and mercury ions will attack aluminum. Even small amounts in contact with the aluminum surface will quickly corrode the surface of aluminum. Mercury ions in solution will rapidly corrode aluminum and anodized aluminum.

Soils

Certain soils can be harmful to aluminum. In particular, clay soils can be most damaging to aluminum if in constant contact. Aluminum immersed in moist clay soils will corrode regardless of the alloy. Protective measures must be utilized to prevent corrosion in clay soils. Test the moistened soil that will be in contact with aluminum to see if the pH is outside the recommended range of 4–8.5. If the soil is outside the range, use a chemical inhibitor coating, such as a heavy layer of bituminous material. A good paint, such as an epoxy, will work, but it should be thick and unbroken. In highly acidic or alkali soils the aluminum will pit and corrode in a very short time.

In dry sandy soils, aluminum will perform well, and no additional precautions are required. In dirt soils that have neutral pH and that are well drained, darkening and staining of the surface will occur, but corrosion will be slow.

Concrete

Wet concrete will attack aluminum and etch the surface. Fresh Portland cement contains lime, which is calcium hydroxide. Mixed with water, the solution will form an alkali. Once the concrete is dry and hardened, it will not affect the surface of aluminum. When casting aluminum into concrete, coat the surface with a heavy bituminous layer of paint or, better still, sleeve the placement and allow the concrete to dry, then place the aluminum. Aluminum imbeds are a major concern to engineers because of the tendency wet concrete has to corrode aluminum. Once the concrete solidifies and air is kept from the aluminum, there is little corrosion that will follow.

Acids

Aluminum will be attacked by some acids but not by others. Nitric acid, for example, will attack aluminum when in a dilute form, while concentrated nitric acid will not harm aluminum. Phosphoric acid in dilute form will not harm aluminum, but in concentrated form it will corrode aluminum. Most organic acids will not harm aluminum while sulfuric acid, hydrochloric acid, and hydrofluoric acid will attack and corrode aluminum.

Alkalis

Most alkalis will harm aluminum surfaces. Lime, which is calcium hydroxide, will corrode aluminum, as will sodium hydroxide and sodium carbonate solutions. Sodium carbonate solutions turn into a weak carbonic acid but a strong sodium hydroxide base. Adding silicates will improve resistance of aluminum to alkali attack. When cleaning with strong alkalis, add silicates to the mixture to protect the aluminum against alkalis with a pH as high as 11.5.

It is important to note that zinc will protect aluminum as a sacrificial anode except when the two metals are exposed to a strong alkali. In alkali solutions, zinc and aluminum reverse potential and the aluminum will corrode.

CLEANING

It is important that the surface of aluminum is kept clean. Buildup of dirt and other substances on the surface can become sites for corrosion to set in. See Table 7.5. The coarser the surface, the more it will retain substances against the self-cleaning contribution of rains. Thus, a smoother surface finish and a periodic wash down with fresh water is advised. For most cases, simply washing down periodically with clean fresh water will remove chlorides and other pollutants. Deionized water is even better. Deionized water will capture the salts and carry them away.

The use of alkaline cleaners and detergents should not be used unless they contain a silicate inhibitor. Strong alkaline cleaners will affect the oxide layer on aluminum. Anodizing, which is a thickening of the oxide layer will also be affected by strong alkaline cleaners.

TABLE 7.5 Substances that will not corrode aluminum and substances that will.

These substances will not corrode aluminum	These substances will corrode aluminum
Ammonium hydroxide	Hydrochloric acid
Acetic acid	Sulfuric acid
Fatty acid	Phosphoric acid
Concentrated nitric acid	Dilute nitric acid
Distilled water	Sea water with heavy metals
Deionized water	Wet wood
Sulfur dioxide	Calcium hydroxide
Hydrogen sulfide	Sodium hydroxide

If corrosion is present, such as the small grayish spots, cleaning should be performed using a more rigorous approach. First use a mild acid, such as vinegar (acetic acid) mixed with cream of tartar (potassium hydrogen tartrate). Make it into a paste and gently rub it on the surface. Aluminum hydroxide is insoluble in solutions above pH of 4.5.

There are several commercial cleaners that work quite well. These are stronger solutions and can be found at automotive parts stores. They are used for cleaning aluminum wheels. These solutions usually contain stronger acids, such as phosphoric acid. Some contain ammonium bifluoride, which is a mild etchant and a precursor to hydrofluoric acid. Exercise caution when working with acids.

To prevent this type of corrosion from occurring, a regularly scheduled cleaning protocol is recommended. Clean the surface with an aqueous solution in the pH range of 4–8.5. Use mild detergent and a strong rinse with clean water, preferably deionized water.

PROTECTIVE MEASURES

The most important first step in determining how to address potential issues with aluminum is to examine the environment in which the metal is to be exposed. Determine the composition and conductivity of the surrounding materials. If the electrical conductivity is low, then the corrosion rate is low. If the surface has good positive drainage, lacks areas where corrosive products can collect, and is able to receive periodic natural washing from rains or from maintenance, the surface should perform well.

There are several ways to protect the aluminum surface and slow the corrosion of the metal. As mentioned, marine grades are best for exposures of aluminum in the environment.

Anodizing the surface can protect the aluminum from this corrosion. Paint coatings can as well. Scratches through these coatings, edges, and fasteners penetrating the surface can be areas where corrosion can develop and initiate. Aluminum does not receive paint well unless it has been treated in some fashion. Phosphatizing the surface and anodized aluminum surfaces are good pretreatments to allow painting to adhere. Chromate pretreatments are still the best for preparing paint.

Aluminum does not afford the same type of sacrificial protection as galvanizing (zinc coating of steel); instead, when a scratch occurs deep into the surface, the oxide layer will rapidly develop and protect the surface. The nature of the gouge may make it prone to hold substances that will affect the aluminum and promote corrosion.

The following is a list of protective measures afforded to aluminum in corrosive environments.

- Use marine grade alloys
- Consider alclad alloys
- Anodize the aluminum surface
- Cathodic protection
- Chemical inhibitors
- Organic coatings

By cathodic protection of aluminum, we wish to transform the aluminum into a cathode. For most installations, aluminum wants to be the anode due to its position on the electromotive force table.[2] So if we can force the aluminum to be a cathode, it will perform well in many corrosive exposures. One means of doing this is to attach a magnesium rod to the aluminum. This requires a periodic replacement of the rod.

Chemical inhibition is another way of isolating the movement of electrons in a galvanic cell. The desire is to insulate the surface and restrict the flow of electrons. One should also avoid grounding an electrical circuit to an aluminum structure. The aluminum's desire to be an anode will increase the rate of corrosion of the aluminum.

PAINT COATINGS

Aluminum is an excellent material as a base for high-quality paints. For the last several decades aluminum has been the base metal for nearly all paints used in industry. The main reason for this lies in the inherent corrosion resistance of the metal. When scratched, the exposed aluminum nearly instantaneously forms a protective barrier. The edge of aluminum sheets sheared to a blank size will not corrode and will resist under-film corrosion.

The major problem is that paint just does not adhere well to bare aluminum. Aluminum surfaces must be pretreated to receive paint. Often called a conversion coating because it converts the surface of aluminum to one that allows paint to adhere when exposed to the environment. Pretreating the aluminum to allow the paint to adhere is critical for long-term success regardless of

---

[2]Electromotive force, EMF, is a voltage measurement of the predicted electrical charge produced by different metals in an electrolyte. Michael Faraday was the first to establish the relationship between two electrodes in an electrolyte and arrived at a measurement for the voltage cell.

FIGURE 7.17 Custom printed aluminum surface. Formed sheet ceiling for "Clydes Wine and Dine," New York. Designed by Morphosis.

the paint system or type used. The NASA Space Shuttle used a chromated primer to protect the A92024 alloy aluminum. NASA engineers incorporated a barrier layer of room-temperature vulcanizing silicone rubber called RTV. RTV is available in different hardness and is used in mold making. Most art and architectural applications do not need to fly through the atmosphere at 20000 miles per hour, however, so powder coatings of polyester or the superior fluorocarbon coatings will work very well (Figure 7.17).

One method to pretreat aluminum is to abrade the surface and then immediately follow with painting. The abrasion removes some of the oxide and adds tooth to the aluminum surface, allowing paint to stick.

Anodizing is another pretreatment that can prepare the surface for painting. The porosity the anodizing process creates on the aluminum surface is excellent for paint adhesion. You will want to put a proprietary aladine coating over the surface to receive paint. The anodized surface is inert and provides a level of corrosion protection in the event that the paint is scratched.

Chemical pretreatments are the preferred method of preparing aluminum for paint. In the past, most of these involved chromate pretreatments that changed the surface to aluminum chromate, which allowed good adhesion of paint. The removal of hazardous chromium compounds from industry has been a major effort. Phosphates and other proprietary treatments have been developed to replace the chromate pretreatments. There are also several developments in pretreatments that have shown great promise in responding to concerns about health issues and toxicity. These involve zirconium, titanium, and silane treatments that bond with the aluminum oxide. There are nanocoatings involving ceramiclike interface with the aluminum oxide surface. These have all shown excellent conversion coatings for aluminum to receive paints.

For aluminum, two important steps are necessary for long-term performance of any paint application: cleaning the surface and preparing the surface with a conversion coating. The paint application performance listed in Table 7.6 assumes these steps have been correctly performed.

Powder coating applies the paint in little beads. The beads stick to the aluminum surface by electrostatic charge. There is no solvent involved and any of the beads that do not adhere to the aluminum surface are recycled. The coating goes on very evenly because of the nature of the electrical charge applied to the beads (Figure 7.18).

There can be a challenge when applying coatings on surfaces that have perforated, cut out, and tight corners. What is referred to as a Faraday effect occurs on these areas due to a repulsive action. A charge cancelation occurs along corners and within shallow openings or holes. These areas lose the charge and the particles will not be attracted.

Once the particles are adhering to the aluminum surfaces, the piece is inserted into an oven where the entire object is heated, and the small paint beads melt and join together to form a continuous layer of paint.

Powder coating can be performed on formed sheet parts, plates, extrusions, and cast forms. It is not used on flat sheets that have yet to undergo fabrication. The process does impart heat into the aluminum form. This can have an effect on large flat surfaces (Figure 7.19 and Figure 7.20).

Coil coating is a very effective means of painting thin aluminum sheet. Maximum thickness of coil-coated aluminum is 3 mm. Beyond that thickness, consider powder coating or liquid spray coating. Coil coating takes the long ribbon of metal and processes it through pretreatment steps and into a final coating applied evenly by a set of rollers. Once the paint has been applied the ribbon of metal proceeds into an oven that heats the sheet up and removes the solvent used to carry the paint. The finish is a superior paint surface with consistent properties. The constraint is in the quantity needed. A full coil of metal is coated at the same time in a single color. The range of colors available is

TABLE 7.6 Common paint application for aluminum.

Paint application	Paint makeup	Performance expectation
Powder coating	Polyester, acrylic	Good to excellent long-term results.
Powder coating	Epoxy, urethane	Good results. Expect fade in 8–10 years.
Coil coating	Fluorocarbon, polyester	Excellent long-term performance. 20-plus years performance.
Liquid spray	Urethane, acrylic	Good. Expect fade in 8 years.

FIGURE 7.18 Powder-coated aluminum plate.
Sun Pavilion, designed by David Dalquist/RDG Dahlquist Art Studio.

Paint Coatings 319

FIGURE 7.19 Custom painted and perforated aluminum for the Miami Design District. Museum Garage. Designed by Terrence Riley and Workac for this portion of the incredible garage façade.

FIGURE 7.20   Fluorocarbon finish on large aluminum trellis panels. Seaside exposure.

extensive, and there are a number of outstanding coil-coating operations hat specialize in applying paint to aluminum.

Once the coil is coated, it is sheared and processed through roll forming operations or general fabrication. The paint coating has the necessary flexibility and adhesion to allow post bending and forming without cracking or delaminating.

Liquid spray coating is similar to powder coating. Liquid spray coating is performed on finish parts, extrusions, and castings. Small batches can be liquid spray coated. The coating is good but not to the level of coil coating. The paint is applied to the surface by spraying from a paint gun. The paint can be electrostatic, similar to powder, but any overspray is waste. Runs, sagging, and differential thickness are common challenges faced with liquid spray applications. These coatings can be air dried or oven baked. The paint is carried to the part in a solvent. The solvent is removed by exposure or heating the part (Figure 7.21).

Paint Coatings

FIGURE 7.21 Wet paint application on large 13-mm-thick plates of aluminum. Polyester. Student Center designed by Legorreta. Doha, Qatar. Jan Hendrix, artist and sculptor for the Wall and Tower.

Paint process	Constraints
Powder coating	Fabricated parts, extrusions, and castings. Small or large quantity. Oven size may limit dimension of piece. No solvent used.
Coil coating	3 mm maximum thickness. Minimum single coil of metal. Limited to coil of unfabricated aluminum. Solvent carrier.
Liquid spray	Fabricated parts, extrusions, and castings. Small or large quantity. Air dry or oven dry. Thickness will vary across the part. Solvent carrier. Waste generated.

Aluminum is soft, ductile, reacts with other metals, and is susceptible to alkali attack. It is critical to keep these characteristics in mind when handling and storing aluminum to avoid unnecessary damage. When handling aluminum sheet or plate, use strong wooden skids, preferably with a plywood base to support the metal and keep it from flexing between slats that are spaced. Nails or screws that stick out of the crate and come in contact with the aluminum will scratch or dent the metal and can impart steel to the aluminum surface.

STORAGE AND HANDLING

Extrusions should be crated in such a way they are prevented from flexing and from rubbing against each other during transit. That goes for cast aluminum parts as well. They should be protected from rubbing against one another, otherwise the surface could be abraded and galling could occur when two surfaces of aluminum rub together. Aluminum, in particular aluminum with a coarse surface finish, is prone to galling if the surfaces of two parts come in contact and enough pressure is applied. Aluminum is usually stored dry and unlubricated. When two surfaces come together, small surface asperities can engage, and the metal will tear. Anodized aluminum surfaces are harder and thus resist galling but can scratch. Separate the aluminum surface with layer of dry paper or foam wrap to keep them from rubbing on one another (Figure 7.22).

Do not allow other metals to come in contact with the aluminum surface. If steel bands are used, then separate them from the aluminum with wood or plastic. When storing aluminum, place them on wood racks rather than steel, or keep the steel contact surfaces covered with wood.

FIGURE 7.22   The importance of correct packing. Isolate each fabricated item to eliminate rubbing and marring.

Moisture should be kept from the aluminum surface during transport and storage. Condensation can develop on the surface in many storage environments when temperature changes abruptly. Unless it is a significant amount and it is allowed to stay on the aluminum surface for an extended period of time, this should not be an issue. If aluminum surfaces do get wet during transit or storage, dry them out. The challenge is that if moisture gets between sheets or rolls of coiled aluminum, a water stain will be created. This doesn't necessarily affect the mechanical properties, but it can be an aesthetic concern. Significant staining issues arise when packing materials absorb moisture and hold it against the aluminum surface for prolonged periods of time. These can actually etch the surface to the degree that subsequent mechanical processes are necessary to return the surface back to an acceptable condition.

CHAPTER 8

Coping with the Unexpected

If you take control over those things you can, you are better able to negotiate the unexpected.

—Judge Judy

INTRODUCTION

All materials used in art and architecture at some point pose challenges in production and then again when exposed to the environment. Art and architectural uses of metal are, for a large part, a subjective effort. How the metal surface appears in different conditions of varying light intensities will weigh on the subjective scale of acceptance or rejection. Ways of mitigating a few of these challenges before they occur or ways to deal with them after they occur will give a bit of understanding to the processes involved with creating and working with aluminum.

All metals are subject to the entropic effects of time. Aluminum is no exception. The more energy put into the refining of the metal from their basic ore, the more stored energy is placed into the metal. Aluminum takes a lot of energy to refine and thus this energy is slowly released as it slowly returns back to the mineral form. This is commonly referred to as corrosion. Corrosive effects are not reversible; however, they can be slowed down to provide a reasonable lifetime. Aluminum has not been around that long compared to other metals, but there are many structures and surfaces created more than 100 years ago that still are performing well. Mechanical properties are still intact. There is a level of surface degradation from the constant interaction with the environment we live in, but other than minor deterioration of the surface, aluminum should look well and perform as designed for years to come.

There are ways in which aluminum will display conditions that differ from the perceived ideal surface. Many of these were there from the beginning when aluminum was first cast and treated to its initial alterations to sheet, plate, and extrusion forms. Others are results from subsequent handling, storage, and manufacture of the metal assembly, while others are induced from outside physical interaction over time with the world at large. How to grapple with these conditions and the potential remedies will be addressed in the following pages.

THE PRINCIPAL OF LIFE CYCLE

Life cycle is often defined as the working lifespan of a product. When it comes to aluminum in art and architecture, there is an additional aspect: appearance. The surface or construct made from aluminum may perform perfectly well but can appear shabby and worn. The aluminum assembly may work as originally expected but may not look aesthetically pleasing to the point that replacement or refinishing may be in order. For example, the doors of the Robert F. Kennedy Department of Justice Building in Washington, DC are made from cast aluminum and aluminum plate (see Figure 8.1). Designed by Milton B. Medary and constructed in 1935, these aluminum doors saw

FIGURE 8.1 Cast and plate doors on the US Treasury building after cleaning.

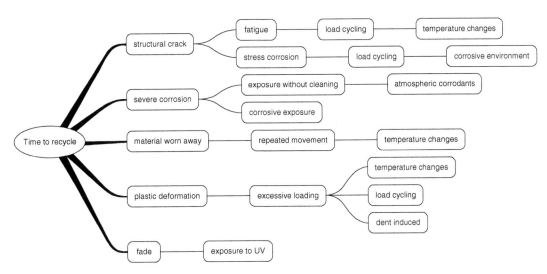

FIGURE 8.2   Various paths leading to recycle of the metal aluminum.

decades of de-icing salts. The surfaces are pitted, but the finish has aged well over the years. Periodic cleaning to remove salts and oxides is all that is needed to prolong the doors' useful life. The finish, however, will continue to weather and age.

Questions that arise when evaluating the life cycle of aluminum include What causes the deterioration? Can it be slowed? and Can it be reversed? Figure 8.2 depicts some of the conditions that lead to the eventual end of the useful life of an object constructed of aluminum. The beauty of aluminum is that after its useful life, it can be recycled and used again.

MILL CONSTRAINTS

The Mill is where aluminum is first produced into the various forms needed for industry. The Mill produces the ingots needed to roll into sheets and plates, ingots for extrusion presses, and ingots for casting. The Mill also casts the ingots that are to be further processed to produce coils or plates.

The market is important to the Mill but the costs of tightening tolerance and the expense of increased risk in falling out of the tolerances is not practical. Architecture and artistic uses of aluminum are only a small piece of the overall market. The building and construction segment amounts to approximately 14% of the market and architectural is but a portion of this. The Mill is going to operate at standards, meet the tolerance requirements, and produce the aluminum in the most cost-effective way to achieve a product acceptable to the marketplace (Figure 8.3).

The industry is influenced by size and need. The Mill producer is working under constraints of scale and industry standards. It will not accommodate special runs or small runs unless they fit within these constraints. It becomes economically challenging to do otherwise.

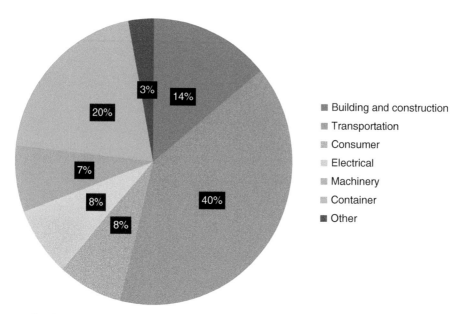

FIGURE 8.3 Aluminum consumption by market.
Source: Data from Aluminum Association 2016.

A Mill first produces a "heat" of metal in a massive furnace. A "heat" is just that, a melting of clean aluminum scrap sorted and identified by alloying constituents and pigs of high purity or special alloying compositions. A heat is a large mass of metal, several thousand kilograms in total.

From the heat, the Mill will produce several ingots. Each ingot weighs from 9,100 to 31,750 kg (20,000–70,000 lbs.). These ingots are cast to a specific alloy for a specific end product such as sheet ingots, extrusion ingots, foundry ingots, and wire. Casting of the ingot is also referred to as a "lot." A lot has a time stamp and date corresponding to the time the alloy is initially melted into the ingot form.

If the quantity needed for a project requires more than one heat, variations on when the casting of the ingots actually occur, will undoubtedly have slight differences in alloying constituents, still within the range of industry tolerances of that alloy. If the metal is rolled into coil material to be later cut into coils, different rolling mills will produce a different surface appearance due to differences in equipment and the surface of the rolls. Different thicknesses and different widths rolled out of ingots from the same heat can also show differences in appearance as the grain of the metal is modified from different processes. When the metal is tempered, particularly when thermal treatments are performed you can expect variations from one tempering treatment on one period of time to that of another. Again, this is due to subtle differences in grain development.

When working with a Mill, acquire a mill certification (see Figure 2.1), and be certain the Mill understands what and how the metal is intended to be used. It is in no one's best interest to provide something that will ultimately tarnish the reputation of the designer, the fabricator, and the metal itself.

POTENTIAL ISSUES WITH SHEET AND PLATE

Lüder Lines

Lüder lines refers to a visual imperfection on the surface of aluminum plates. It occurs when the plates are flattened and stretching occurs. Lüder lines can appear in sheet, as well. These lines are local plastic deformations that result from nonuniform yielding of the metal. This malady can be caused from differential cooling of the plate or from stretching to achieve flatness. They manifest as visual surface lines running across the width of the plate. There are several lines on a given plate and they are parallel to one another like a wave. They extend across the plate and are oriented at approximately 60° angles. Lüder lines can also appear as stains on the surface like flowing moisture and occur from one edge into the plate.

A95xxx aluminum alloy plate is more susceptible to Lüder lines than other alloys. Lüder lines can occur in all alloys but are generally limited to thicker plate material, particularly in the A95xxx series of aluminum.

There are two main types of Lüder lines, distinguished by their regularity across a surface. One being more irregular and the other appearing as parallel stripes, often with an opposing set of strips on the same sheet.

- Type A: Characterized by irregular marks running across the plate
- Type B: Characterized by regular parallel lines running across the sheet at approximately a 60° angle

Lüder lines are aesthetic maladies that are very visible in uncoated plate material. Painting and anodizing do not fully conceal them. They will show through high gloss paint. Glass bead blasting seems to mask them as does surface grinding and polishing. This phenomena are difficult to control at the Mill. Changing the alloy to A96xxx can reduce their occurrence. A6xxx alloy is less prone to the development of the surface imperfection (Figure 8.4).

Lüder lines do not occur often in thin cold rolled sheet but are common in hot rolled and flattened plate material of thicknesses 9 mm (0.375 in.) or greater. They are a visual defect that is in the material rather than only on the surface. You cannot feel the imperfection, but it is very visible. They do not affect the mechanical properties of the aluminum nor do they have an effect on the corrosion characteristic of the metal.

If you are designing with aluminum plate, consider using A96xxx alloy and inform the Mill source of the need to reduce the occurrence of these lines. They are not a defect in the mechanical sense, but they may be aesthetically undesirable. It is important to be aware of them.

Mill Finish Stains

There are several stains that can be seen on newly received aluminum sheet or plate. These can be composed on nonuniform stains from water or oil, heat treatment stains from a contamination in the

FIGURE 8.4 Visible Lüder lines on aluminum plate. Note the angle and parallel sequence in the reflection.

heating oven, and rolling streaks. The subsequent processes performed on the metal will determine if the material is usable. If mechanical abrasion is followed by paint application, then some of these will not matter because they do not impede mechanical properties or corrosion resistance of the base material. If the surface is to be anodized, these stains may translate through the finish. Test a sample to see if the etch removes the appearance. Sometimes the anodizing process will make them appear where they were not visible before. As the oxide grows in different ways than the surrounding oxide, the stain is visually enhanced. This is particularly insidious. All the efforts and cost have gone into the metal only to have the anodizing performed at the end show the defect.

Sheet and Plate Distortions

In sheet and thin plate material that has been reduced and recoiled, there are several quality concerns that should be addressed with the supplier of the base metal. See Table 8.1.

TABLE 8.1 Quality concerns from original metal supplier.

Malady	Description	Recommendation
Bow	Longitudinal and transverse	Leveling
Buckle	Arbor, center, edge, quarter	Leveling
Oil canning	Stress trapped in thin sheets	Leveling, thicker sheet
Coil set	Differential and reversed	Leveling
Crease	Poor handling produces a "fish eye"	Reject sheet
Mars	Long dark streaks on the surface	Reject sheet, mechanically abrade, or paint
Scratches	Deep or shallow gouges into surface	Reject, mechanically abrade
Edge belled	Differential stress causing hour glass	Shear down, level, reject
Orange peel	Granular surface structure	Reject
Roping	Usually after severe deformation	Smooth out, reject
Structural streaking	Nonuniform, often banded appearance on surface.	Residual from casting. Not repairable.

These maladies can be overcome in some instances, and in others the material should be rejected and replaced. Stretcher leveling of the thin sheet metal when it is decoiled is the most direct way of improving the appearance and removing noncompliant distortions from the surface. Creases, mars, and scratches in the surface are due to mis-handling of the metal and should be rejected if they cannot be abraded to cover or if they cannot be cut out of the finish blank.

Damage can also occur at the fabrication facility where the sheet, plate, extrusion, bar, or tube are worked into the final design elements. See Table 8.2. Most of these are due to improper storage and handling of the material as it flows through a fabrication facility and is delivered to the final installation.

Scratches from mishandling aluminum in the fabrication facility is a sign of inexperience or lack of care. Both are not acceptable to a designer using the metal. Scratches add work in order to repair or restore the surface. Paint coatings and anodized coatings will not conceal the scratches. Protect the metal at all times as it moves through the factory and on to finishing. Scratches in finish goods that occur during shipping if parts are allowed to rub against one another or other surfaces can ruin the aluminum work.

"Fisheyes" are another sign of inexperience and lack of care. These are small creases where the aluminum has undergone permanent plastic deformation. They occur when thin sheet material is carried improperly and the metal yields in the center.

If steel particles become embedded in the surface, they can create corrosion cells later on exposure as the surrounding aluminum becomes sacrificial to the steel. They are difficult to see until exposure to a wet environment. If the subject part is to be anodized, an acid cleaning treatment

332 Chapter 8 Coping with the Unexpected

TABLE 8.2 Quality concerns created at the fabrication facility.

Malady	Description	Recommendation
Scratches	Induced into the surface by dragging over harder substances	Protect the table. PVC protection on metal during fabrication.
"Fish-eyes"	Small dents from improper handling of sheet material.	Permanent deformation. Cut around or reject.
Embedded steel	Airborne steel particles or deposits from scratching introduced to the surface and become embedded during fabrication.	Difficult to see until exposed to weathering. Isolate equipment to aluminum only. Separate steel fabrication. PVC protection of surface.
Mars	Streaks on surface created from dragging an aluminum part over another aluminum part.	Protect with PVC. Avoid dragging one piece over the other. Apply a finish that conceals streaks.

may be required before etching in sodium hydroxide. The anodizing process won't remove these steel particles.

Mars occur in mill finish aluminum when one surface of aluminum is dragged over another. A dark streak results. The steak cannot be wiped from the surface but must be polished out.

Many of these can be eliminated if the metal is properly handled and if the aluminum has a protective film over the surface as it is being worked into the final forms.

DENTS AND SCRATCHES IN ANODIZED AND PAINTED ALUMINUM

Aluminum is a soft metal, softer than stainless steel. Its lack of hardness and ductility make it easier to damage from mishandling and exposure to abrasive contact. This characteristic, though, also makes it easier to repair and reverse the damage. Scratches can be polished out. Dents that have plastically deformed the aluminum can be reduced or removed.

Anodized surfaces, however, are significantly harder. These surfaces lack ductility. Once aluminum is anodized, it cannot be postformed without fracturing the thin, ceramic-like oxide. A sharp blow can dent the anodized aluminum and, in the process, create small cracks around the dent as the inelastic anodic film fractures. There is no way to reverse this. The indentation from the blow may be reversible but in doing so, more small cracks in the anodic film will appear. Scratches through this film are also not reversible. The anodic film is more resistant to scratching, but if it is damaged, it cannot be repaired unless the anodic film is removed and the scratch polished out, then restoring the film by reanodizing the surface.

Piercing, shearing, and perforating anodized aluminum will induce small microcracks around the cut. This is due to the inelastic nature of the thickened oxide layer. As the underlying metal begins to yield, elongation occurs, which fractures the inflexible thickened oxide (Figure 8.5).

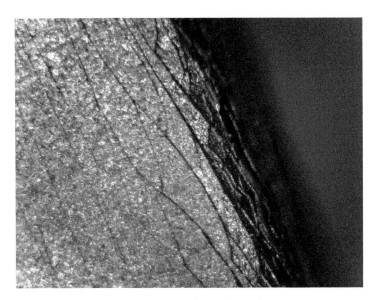

FIGURE 8.5 Microscopic image of fractures on anodized aluminum sheet.

The thicker the base aluminum is, the more that this fracturing will be apparent. Thicknesses greater than 1 mm (0.040 in.) are most susceptible to the microcracking. On thin anodized aluminum sheet, the fracturing is less apparent. On dark anodized colors, the fracturing will be more visible and will appear as light contrasting fissures against the dark background.

Thicker sheet or plate aluminum can show similar cracking conditions when formed, pierced, or sheared. It depends on the flexibility of the paint finish and often the primer or pretreatment layer. Small fissures along the bend line or around the sheared edge will be apparent, particularly on darker finishes. With baked-on enamels, this can sometimes lead to delamination along the bend line or partial delamination at the edge where the metal has been sheared (Figure 8.6).

GRAFFITI

In both art and architecture there is the occasion where some hooligan decides they wish to express an opinion by spray painting some ill-conceived slogan on the surface of the aluminum. Most of the paints used on aluminum are thermoset polymers. By their nature, these coatings are very durable and resist chemical and atmospheric decay. Old paint finishes may have some micro-coarseness to the surface but still possess resistance to mild solvents (Figure 8.7).

Anodized aluminum when sealed properly is a very corrosion-resistant and chemical-resistant surface. Most mild solvents, particularly at temperatures well below the boiling point, will have little effect on the anodized surface or the seal. Mill finish aluminum, depending on the finish, offers little for the spray paint to adhere to and also is resistant to chemical and solvents.

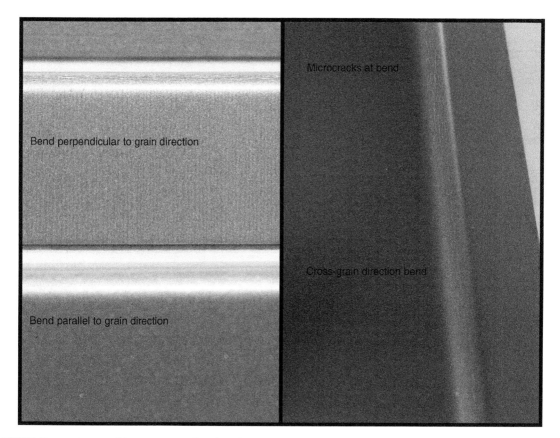

FIGURE 8.6 Image of fractures on painted surfaces.

To remove the graffiti, begin with household solvents such as xylene. This will remove most air-drying paint from the surfaces of aluminum. Thermoset paints will not be damaged if the solvent is wiped on and wiped off, then thoroughly rinsed. Anodized aluminum surfaces will not be affected by the solvent and the offensive paint should come off. Dye anodized surfaces are a little more sensitive because the dye is near the pore outlet. Test an area first to be sure the dye is not affected by the solvent. If it is, you will need to back down to a milder solvent and see if the paint will be removed.

Mill finish aluminum should clean easily and no remnant of the offensive graffiti should be apparent. Do not use abrasives to remove the graffiti.

FLATNESS

For sheet and plate material used in art and architecture, flatness criteria are an important constraint. For aluminum, this is not as troublesome as other metals with high internal stresses induced from cold rolling processes. Still, nonflat surfaces can be present in metal that has been stored improperly or cold rolled on equipment at the Mill that has not been calibrated properly.

Minor variations in and out of plane will be apparent in different lighting conditions.

Attempts at establishing acceptable criteria should be discussed ahead of time with the metal supplier and end user.

One measure used to establish flatness criteria is the formula:

$$I = (\pi/2 \times H/L)^2 (10)^5$$

I = Measure of differential length between longitudinal elements of a rolled product.
H = Maximum height (amplitude) of an out-of-flat section, valley to peak.
L = Average straight-line wavelength of an out-of-flat section, peak to peak.

This formula is when there is shape across a single element and is more for measuring conditions depicted in *Condition B*, of Figure 8.8. This condition may manifest when stiffeners are placed across the surface.

The I value should be no greater than 0.002 for a measure of flatness on reflective or glossy surfaces, which is tight but should be possible to achieve. If it falls out of this maximum, then a prototype and visual review should be undertaken to be sure all parties agree. If more relaxed values can be considered, this should come out of the prototype review.

Condition A is a common condition known as pillowing. In this condition, element to abutting element may create a condition that is not desirable. Usually if all the pillowing is in one direction, the surface will look more acceptable. If they pillow in and out, the aesthetic appearance will be significantly different. Also, if the pillowing is severe, flatness is no longer achieved, and you have an undulating surface. For pillowed elements consider using H/L alone since wavelength is element to element rather than on a single element. Here H/L should not be greater than 0.0015 (Figure 8.8).

Roll forming operations can lock in distortions that are in sheets or coils. Leveling operations may assist in removing some of them. Break forming operations can also lock in internal stresses, so beginning with flat sheets is critically important.

So, what are acceptable levels of out-of-flatness? This is dependent on the use and how reflective the finish is, but the following are some constraints to consider.

1. *Thickness*. Thicker sheet will give a flatter appearance.
2. *Thermal movement*. Design how a surface will expand and contract.
3. *Reflectivity*. More specular surfaces will exaggerate minor differences.
4. *Viewing angle*. Grazing angles show minor differences in flatness.

FIGURE 8.7 Coil-anodized aluminum surface test removal of graffiti. Paint had been left on the surface for over 24 hours before it was removed.

A good process will involve discussions of these with the fabricator, end user, and designer in advance and arrive at a solution that is acceptable to all. A prototype of the actual, worst case may be warranted to arrive at a level of visual acceptance.

There is no substitution for thickness simply means that there are physical behaviors in thin plate diaphragms that cannot be overcome without increasing the stiffness of the material. Stiffness can be increased by adding section such as ribs or laminating thicker materials to the reverse side or by increasing the thickness of the metal. Thermal movement can play into this as well. Aluminum has a coefficient of thermal expansion of approximately 23×10^{-6} m $(mK)^{-1}$ depending on the alloy. In comparison to other metals and other materials this is high. This means it will expand and contract when it is subjected to temperature changes. See Table 8.3.

Where this has an effect on aluminum is the ability to "push" the thin plate away from a given fixed point without visible distortion. The expansion, or contraction pushes material away or pulls toward a point of fixity. If the section of the metal is not sufficient it will create a surface distortion. Some refer to this as a "buckle," but this is not the case. A buckle is a permanent plastic deformation in a surface. The surface distortion is not permanent and as the temperature changes the occurrence and appearance changes (Figure 8.9).

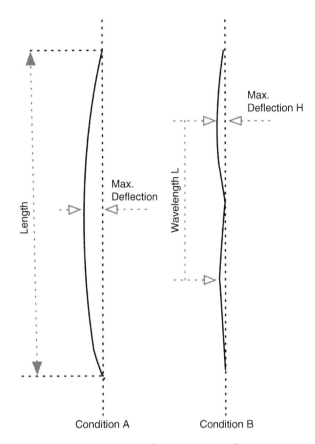

FIGURE 8.8 Methods of establishing measurements for determining flatness.

TABLE 8.3 Approximate coefficient of thermal expansion (m (mK)$^{-1}$).

Metal/Material	Coefficient of thermal expansion
Aluminum	23×10^{-6}
Copper	17×10^{-6}
Steel	12×10^{-6}
304 Stainless Steel	17×10^{-6}
316 Stainless Steel	16×10^{-6}
Titanium	9×10^{-6}
Zinc	23–33×10^{-6}
Glass (plate)	9×10^{-6}

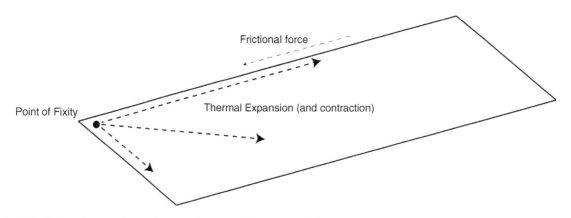

FIGURE 8.9 Contraction and expansion to and from point of fixity.

This is also known as the columnar strength of the material. The thin sheet or plate has to act as a column and overcome the forces internal to the sheet and the external frictional forces that may be playing into the resistance. Think of a thin piece of paper: the larger the paper, the larger the diaphragm. To move the paper by pushing from the edge works fine until the paper gets larger, which increases the frictional forces. If you crease the paper, it will increase the section and strengthen the diaphragm. If you used cardboard or thicker paper it will also resist distorting because of the increased stiffness that thickness affords.

To determine how much thermal expansion to expect or to design for, use the following formula with the table of coefficient of thermal expansion.

$$\Delta L = L_i \times \partial (t_f - t_i)$$

ΔL = Change in length expected
L_i = Initial length of part
∂ = Coefficient of thermal expansion (for aluminum use 0.000 023)
t_f = Maximum design temperature
t_i = Initial design temperature

For example, if the temperature in the plant fabricating a 3000-mm-long aluminum assembly is 10°C (50°F) and the part is going to be installed in a hot place like Qatar where the temperatures will reach 55°C, you can expect the following change in length.

$$\Delta L = 3000 \times 0.000\ 023\ (55 - 10)$$

$$\Delta L = 3.105 \text{ mm}$$

The design must allow for the movement of 3.105 mm.

Similar with fabricating parts of aluminum in warm areas or during the summer months when the metal is warmest, you must allow for the contraction of the aluminum. Darker surfaces will get considerably hotter than the surrounding air.

Reflective surfaces show the distortions from thermal movements and internal stress more readily than less reflective surfaces. As the metal wants to push or pull from thermal changes, the surface may appear with warp in flat open regions. Light-scattering matte surfaces can sometimes accentuate the apparent rippling on the surface as well. This is due to localized contrasting dark and light areas that manifest to the viewer from changes in reflection.

One approach is to consider the width to thickness ratio of a large part or surface. If the width between stiffener ribs or edges is considered in relation to thickness of the material and finish, a design could consider the following.

Finish	Width-to-thickness ratio
Highly reflective	150 maximum
Medium to low	200 maximum
Textured or embossed	200 and higher

These can be conservative. It is advised to make a prototype with the finish and at the width and thickness that will represent the appropriate visual appearance.

Note that *visual flatness* is an arbitrary term that is very subjective and should never be a determining factor in whether something meets or does not meet a project's requirements. It is unmeasurable. It is one of those conditions where "you know it when you see it." Viewing angle, time of day when the surface is viewed, and viewing distance are starting places to box in the visual flatness criteria. If this is the concern, then go back to the first point: thickness. There is no substitute for this quality in aluminum. Thickness equates to more material and thus more cost. Additions of stiffeners and laminating thickened materials will add durability but need to be reviewed as follows:

- Viewing angle (s)
- Time of day (sun intensity and temperature)
- Viewing distance

If these are not established as criteria, then the chances of achieving some level of satisfaction go down. The problem is that, when installed on the finish surface, these conditions become altered by the reality of the physical space that the surface ultimately occupies.

STIFFENER SHOW-THROUGH

The approach to strengthen thin sheet metal surfaces often is to add stiffeners to brace the diaphragm and increase section properties across the flat expanse. Stiffeners can be extruded angles, channels, or custom shapes. These stiffeners can be applied with adhesives or fusion studs. Aluminum receives both well if properly prepared. Adhesives can be silicone or epoxy, they can also be dual faced adhesive tapes sometimes referred to as VHB tape for very high bond. One aspect of the adhesives is that they adhere very well to clean aluminum and they possess mechanical properties that meet engineering requirements.

The fusion studs will work well with aluminum as described in Chapter 6. The challenge is to apply the stud to the reverse side of a sheet so as not to make it apparent on the finish face side. The addition of adhesives and VHB tapes can augment the stud weld and reduce the number needed. Again, there is no substitute for thickness with using either studs or adhesives. If a sheet is too thin, the stiffeners will show on the face surface or the distortions will appear where the adhesive has cured and created a distortion that translates to the surface. Sometimes these distortions cannot be measured with any but the most sophisticated devices, but they are visible. Stud welding can telegraph through to the other side in certain lighting conditions and viewing angles. Studs can also induce deformation to the face side if they have bending loads applied. The soft aluminum bends and a crease is visible on the face surface.

When using stiffeners, they need to be applied with care. They need to be set evenly, both in relation to the sheet but also in the force of application. Curing of the adhesive must be done in a controlled setting and reviewed afterward in various lighting conditions and viewing angles.

If studs show through to the face side, be certain the settings on the capacitor discharge device are correct. Run tests on a similar metal and thickness. Often the choice is to turn down the discharge voltage, but this can lead to a cold fused surface. Rubbing the stud onto an abrasive pad will remove some of the oxide on the bottom of the aluminum stud; however, be careful not to remove the ignition tip. Another consideration is to move the ground closer to the stud weld operation.

Visual flatness is an arbitrary term that is very subjective and should never be a determining factor in whether something meets or does not meet a project's requirements. It is unmeasurable.

ISSUES ENCOUNTERED WITH CASTING ALUMINUM

Aluminum is prone to collecting impurities during the casting process. This is due to the specific gravity of aluminum. Many impurities have specific gravities that are near aluminum, so in the molten state, aluminum tends to attract them. In particular oxides are quickly drawn to aluminum. Aluminum is so attracted to oxygen that even in the molten state an oxide will quickly form on the top surface. This dusty gray layer is composed of impurities and oxides and must be kept from the metal during casting. However, this layer, known as dross, does serve a valuable purpose. When it forms, it creates an unbroken layer of oxides that keep the balance of the metal below the oxide free of contamination.

Molten aluminum will absorb hydrogen readily. The hydrogen comes from the air, from the mold, or from combustion gasses present in the facility. Hydrogen absorption will lead to porosity in the casting. Pores can occur within the casting or as pinholes on the surface. Absorption of hydrogen will increase as the temperature of the metal increases; therefore, during the molten stage, absorption is the greatest concern. It will also increase as humidity in cast facility increases. High humidity and moisture will feed hydrogen into the molten metal during the time that the aluminum remains in the molten state. To reduce the absorption of gases, preheat the scrap and feed metal to around 480 °C (900 °F) to remove moisture from the material prior to bringing the temperature up to casting temperatures. Do not leave the metal in the molten state for long periods of time. Most casting facilities bring the aluminum up to melting just prior to making the pour.

Aluminum casting is prone to a challenge for all cast metals known as "hot shortening." When the aluminum is being melted, separation of alloying elements can occur and collect around grain boundaries when solidification begins. Aluminum has very low strength at the temperature just below the solidification temperature range. Shrinkage occurs and this creates a failure crack in the semisolid casting. This is an irreversible condition and the cast part must be discarded and re-melted.

Appropriate design for shrinkage must be made in the mold used for casting aluminum. Aluminum undergoes significant contraction due to the cooling of the metal. Aluminum can shrink in volume from 3.5% to 8.0%. So, the mold, feed, and riser design are critical when casting aluminum, regardless of the alloy.

ISSUES ENCOUNTERED WITH ALUMINUM EXTRUSIONS

Structural Streaks

Aluminum extrusions are linear products produced by pushing a heated billed of aluminum through a hardened die. On occasion a contrasting dark streak will appear, running the length of the extruded form. The contrast is not drastic but can, in some instances, be apparent. It is usually not visible until the aluminum has been anodized.

The streak is created from elements in the original casting that were not sufficiently dispersed or that have collected in such a way they get spread out as the metal undergoes hot rolling or extruding (Figure 8.10).

Other Flaws in Extrusions

On rare occasions oil can be present when the extrusion is pushed through the die. The oil can have an effect on the extruded form by creating an internal cavity. This occurs when they continuously load billets. Some lubrication may get trapped between the billets. As the extrusion is pushed this can cause a separation within the aluminum.

This type of flaw is very difficult to uncover and can appear as a small surface protrusion or bump (Figure 8.11).

342 Chapter 8 Coping with the Unexpected

FIGURE 8.10 Streaks along the length of extrusions.

FIGURE 8.11 Lubricating oil creating internal separation of the aluminum.

Forming Maladies

Aluminum is soft, in particular relative to steel dies. Any imperfections in the dies will translate to the aluminum surface. In brake forming operations, the aluminum is shaped by metal dies under pressure. As the die comes into contact with the aluminum surface, the aluminum goes into compression and tension from one side of the thin plate to the other. There is slippage that will occur as one surface moves against another. This can impart mars or scratches into the softer aluminum surface. The skilled press-brake operator will use protective covers on the dies to keep from marring the metal being formed. These covers do not interfere with the forming but protect the metal surface and allow it to slip without marring the surface.

Marks can be imparted to the aluminum at the form lines when the tool or die used to shape the aluminum element has small nicks or impressions. The pressure is sufficient to permanently set these nicks into the formed aluminum surfaces. The tools and dies are made of hardened steel and should be kept in pristine condition. If a tool or die is damaged it should be taken out of service, refinished, or scrapped.

Uneven bend radii from one section to another are not because of uneven internal forces along a sheet as much as it is in improper settings on the brake. Power brakes have compensators built in to accommodate the differing pressures along their length. If the compensators are not accurately set, the radii will be different along the length. Two parts designed to butt into one another must be made and formed precisely the same, otherwise you will have offsets at the junction of the two components. This can occur with uneven forming or equipment that is out of alignment.

Challenges to Anodizing Aluminum

The process of anodizing is a conversion process, not a coating or a deposition of one material over another. It will not hide irregularities in the original aluminum surface; instead, it may enhance them. The etching process that proceeds the actual anodizing dulls the surface, which can hide some very minor surface irregularities, but it will not remove scratches or smears in the surface, nor will it remove organic contamination such as fingerprints, grease, and oils.

Scratches or blemishes in a sheet of aluminum to be anodized can be polished out but the entire surface will need to be polished, otherwise the polished portion will appear slightly different as light passing through the clear film reflects off the modified finish. The anodized surface is created by growing in and out of the aluminum as it stands. If this is in an element of an overall surface, every element may need to receive the same polish treatment if the desire is a monolithic appearance. If not, the one altered surface will stand out.

Anodized surfaces can be stripped and reprocessed; however, they may not match adjoining elements. Stripping is re-etching the surface by re-emersion in the sodium hydroxide etching bath, but in doing so, some of the oxide usually remains. In addition, the thickness will be reduced. A small portion of the surface is removed each time the surface is etched.

Glass-bead blasting the surface—that is, "Angel Hair" finishing the surface with a very fine grit to even out the appearance—can prepare the aluminum to be effectively anodized. If, however, you are doing this to remove a scratch or even out an aluminum surface, it will not match adjoining elements unless you perform this same treatment on all the adjoining surfaces. You will want to test the treatment and determine if there is a tone difference.

The following are a few of the issues faced with the electrolytic anodize process and the probable cause. Basically, there are three categories that generate the issues occasionally seen when an anodized surface is used. See Table 8.4.

1. Production process of the wrought form of aluminum
2. Production process of the anodizing plant
3. External/environmental effects

Another important part of the anodizing is the sealing of the surface. This is the last step in the anodizing process. There are three methods used most often in the sealing of the anodizing process.

1. Hydrothermic sealing involves immersion of the aluminum anodized product in a hot bath of water. The bath temperature is kept at 98°C (210°F) and the pH is stabilized at a range of 6–6.4. This bath will convert the aluminum oxide to a hydrated aluminum oxide ($Al_2O_3 \cdot H_2O$) at the surface. This larger molecule will affectively seal the pore.
2. Nickel Acetate sealing involves immersion in a heated bath containing hydrothermic sealing but at a lower temperature, of 79–85°C (175–185°F), and with the addition of the larger nickel molecule precipitated into the pores of the metal. This is the most common sealing process used.
3. The cold sealing process also involves nickel in the bath. The temperature is significantly lower, 32–34°C (90–94°F) and the nickel salt reacts with the aluminum oxide surface to seal the pores.

Sealing is critical for long-term performance of the anodized coating. Once the aluminum is removed from the sealing bath it is immersed and rinsed in fresh water to remove excess seal residue from the surface. There may be a visible film on the surface. Wiping the parts down and letting them dry thoroughly usually eliminates this film residue.

CRAZING OF THE SURFACE

As noted, the anodizing process creates a thickened oxide on the surface. This hard layer is ceramic-like in nature. It lacks flexibility. It will not stretch or compress. This inflexible surface layer will crack when formed. Depending on the thickness, the crack—or, as it is often referred to, the crazing—of the surface can be significant. The underlying aluminum will bend but the outer surface will have numerous small cracks extending along the bend and sometimes out from the bend.

TABLE 8.4 Examples of issues and source of potential problems when anodizing.

Malady	Reason		Probable cause
Dark edges or ends	Uneven coating	Anodizing plant	Incorrect voltage on power supply. Bad part distribution on rack.
Light edges or ends	Uneven coating	Anodizing plant	Incorrect voltage on power supply. Bad part distribution on rack.
Nonuniform color	Uneven depth	Anodizing plant	Incorrect ramp up in programed electricity.
Nonuniform color	Uneven depth	Anodizing plant	Bath chemistry incorrect.
Wrong color shade	Uneven depth	Anodizing plant	Incorrect bath temperature.
Burn marks on surface	Current density to great	Anodizing plant	Bad or uneven contacts used in the anodizing process or too fast a ramp-up on power supply.
Pitting of the surface	Contamination	Anodizing plant	Contamination in bath.
Spalling of surface anodizing	Film detachment	Anodizing plant	Too high a voltage.
Nonanodized spot on surface	Electrode	Anodizing plant	The electrode used to pass the current through the aluminum is attached to a visible element.
Stains in anodized film	Alloy elements on the surface	Anodizing plant	De-smutting operation was not thorough and left elements on the surface.
Small lacelike cracks on the surface	Thermal changes or bending	Anodizing plant	Quick temperature changes in rinse process or mishandling of the parts.
Corrosion on surface after rinse	Corrosion of the surface	Anodizing plant	Aggressive rinse water or extended time in rinse tank.
Drips and runs	Fluid running out of fixture or recesses	Anodizing plant	Etchant or acid trapped in the piece being anodized or dripping from fixtures used to hold the parts being anodized.
Comet trails on surface	Contamination	Production process	Polishing compound left on the surface of the aluminum.
Longitudinal weld visible—extrusions	Inclusions of oxides	Production process	Becomes more visible after etching process. Formed at junction of metal joined as it exits the ports on the extrusion die

TABLE 8.4 (*continued*)

Malady	Reason		Probable cause
Transverse weld visible—extrusions Dark V-shape mark	Oxidation discoloration	Production-process	Created by the oxidation on the lead in of the extrusion. Unavoidable. Trim the extrusion end.
Color differences	Inconsistency in alloy	Production process	Wrought material has slight differences in alloying constituents.
Color differences	Subtle differences in the surface	Lighting	Unavoidable. Energy of the light source interplays with the surface of the metal, making some look lighter than others from various points of view.

Additionally, when curving anodized aluminum plates or sheets, you can hear a crushing sound as the anodized surface cracks as it is stretched or compressed.

From a corrosion standpoint, the crazing does create areas where impurities can reside, but the exposed aluminum base metal will rapidly form the oxide, just not the same thickness as the surrounding anodized surface.

On certain surfaces, you can fill the tiny cracks with enamel to blend with the color. On clear anodized surfaces, the cracks will appear white against the darker gray tone background. On colored surfaces, the white cracks will stand out, often feathering out from the bend.

The only way around this behavior is to postanodize the fabricated part. There is no postanodizing repair that can be performed other than the filling of the small cracks with paint or silicone. You would wipe it on and wipe off all residual, leaving the paint in the tiny cracks.

Color Differences in Anodized Aluminum Surfaces

Anodizing is a beautiful surface on aluminum; however, matching color and tonal appearance from one batch to the next will not be exact and can show significant differences in different lighting conditions. Anodizing is a conversion coating, and as such is not a coating on aluminum that covers or provides pigmentation unless it is dyed.

You can expect the appearance of two elements of aluminum to match when anodized if they are taken from the same coil or extrusion billet of aluminum, if they are anodized in the same run, in the same tanks and in sequential times. If not, you will experience color and tone differences.

Different alloys will appear differently when anodized.

Different heats of the same alloy can appear differently when anodized.

Different tempers of the same alloy will appear differently when anodized.

Different time periods can affect the results of clear anodized because the electrochemical process may vary slightly.

The electrochemical process will react to the aluminum alloy surface at slightly different rates and produce differences in film clarity that will have an effect on how the surface appears in various lighting conditions.

CAUSES OF TONAL DIFFERENCES

Appearance and color variations are common occurrences in anodic films created for art and architectural applications. Expectations of perfect color consistency is not going to be achieved in most anodized surfaces because the variables are too numerous and diverse (see Table 8.5). It is not a coating like paint with a mix of pigments and dyes to produce color, but a clear oxide that grows out of and into the surface of the aluminum as electrochemical action interacts with the metal's surface.

The anodized surface variables that can produce different appearances and tones are the coating thickness, the salt used to arrive at the color, tin or cobalt salts, process, either batch or coil anodizing, temperature, solution concentration, type of etching, cleaning and degreasing, time, method of racking, and current density.

The anodized surface used most often today is created in the sulfuric acid electrolyte process for clear and for color in the two-step anodizing process. Appearance in either case is affected by the clarity of the clear oxide film. Appearance differences are more than simply color. Gloss difference and textures, both micro and macro, play a role in what one sees in an anodic surface.

Clarity of the thin film is affected by the alloying constituents at the surface. During the anodizing process the aluminum is initially etched in a strong alkali bath, usually dilute sodium hydroxide. This alkali will dissolve the aluminum on the surface. Since all the aluminum used in art and architecture is composed of alloys, as the aluminum portion is dissolved at the surface, the other alloying

TABLE 8.5 Variables to overcome when working with anodized aluminum.

Variables At the mill source	Variables At the anodize facility
Mill practices	Thickness of oxide developed
Consecutive ingots	Process—two-step, dye, integral
Date of heat	Metal salt used: tin or cobalt
Rolling mill differences	Type of two-step solution
Different widths	Temperature and concentration of solution
Different thicknesses of material	Etch used
Temper	Process time
Flow lines	Racking and size of load in tank
Rolling lines	Current density

elements remain. When the oxide is developed in the anodizing step, these alloying elements mix with the clear oxide as it develops and can have a consequence on the tonal appearance. Trace amounts of iron, for example, will cause the clear oxide to be a shade darker than aluminum of the same alloy with less iron. Therefore, in different heats, there will be differing amounts of intermetallics on the surface of the etched aluminum. These differences will affect the relative clarity of the clear oxide that grows onto and into the surface of aluminum. In the two-step process, color is derived from light interference and from the finely divided metal salts deposited at the base of the pore. Because of this, light reflecting off multiple surfaces back to the viewer is the method that creates the color perceived. There is a marked angularity to this perception that can vary from one element in a field to another. Fading or disturbances in the appearance develop over years of exposure as the clear oxide film roughens. The once intense colors begin to lose some of their tone as the surface ages. Because they are mineral deposits within the pores, they do not have pigments that can be affected by the ultraviolet light. It is the roughening of the outer surface that dulls the reflection.

RANGE SAMPLES

Range samples are often the control device used to establish whether a surface is acceptable or not. Range samples are typically considered extreme allowable dark tones and light tones. This may be an adequate approach to paint coatings, but it falls short as a control devise for anodized aluminum. First, the range samples may have been produced on metal from a different heat than what will be used to create the surface. Second, the range samples were created at a different time in a different bath with a slightly different chemistry.

Range samples will not take into consideration the angular appearance, the gloss, texture, and the light energy level where the subject control is to be established. In bright sunlight, the product may be fine and fall within the range, while in shadow it will not. Range samples are the expected degree of variation that can occur. They are not "the variation" that will occur. If the intention is to place boundaries, both light and dark extremes, objective criteria will need to be established; otherwise, the appearance is up to visual subjectivity.

Objective data along with subjective evaluation is needed to determine the viability of the finish. Colorimeters can establish a color plot. Most anodizing facilities make use of a colorimeter. The colorimeter will provide a value in one of several categories. One of the most common is CIE LAB,[1] which establishes three color values for comparison. From these values one would establish a delta E value.

For example, the CIE LAB would give the values L, A, B. The A and B values can be positive or negative. To arrive at a Delta E you would read the values from the colorimeter readings of the two "range" samples and plug them into the following formula.

$$\Delta E = \sqrt{(L1 - L2)^2 + (A1 - A2)^2 + (B1 - B2)^2}$$

---

[1] The CIE LAB was introduced by the International Commission on Illuminumation (CIE). It provides color as represented by three letters: L for lightness, A for green/red, and B for blue/yellow. These are the components of color.

Delta E values are from the printing and graphic arts world. A Delta E of 1 for example, is the smallest level of color difference between two surfaces that a human eye can pick up. It is highly dependent on illumination and light energy. For metals, angularity also plays a role in how well a surface will match in color. A Delta E of between 3 and 6 is common and generous.

Gloss meter can measure the gloss. Using a 60° relative gloss meter will aid in determining the gloss within some range. Both color and gloss must be judged in a range. Texture and angle of view are more difficult, as is light energy. Daylighting at different times of the day and different times of the year can affect the tone seen in a natural metal surface. When viewed in shadow or indoors, aluminum's appearance can be affected by the energy levels hitting the surface and the differing reflections. It is very important for the success of the project that the variables in anodized aluminum are fully understood and the expectations are clearly defined. If an absolute, perfect match, in every lighting condition and over multiple runs, is demanded, then anodizing is not going to achieve this. If, however, you want a deep, beautiful metallic finish that projects a mildly changing crystalline beauty, then anodized aluminum surfaces are the material you should consider. Coil anodized surfaces will achieve a good match in color over a large surface. Just realize that thicknesses obtained with coil anodizing are significantly less than with batch anodizing, and the expectations for long-term performance of the anodized surface will be diminished (Figure 8.12).

FIGURE 8.12 Examples of range samples and the extent of variation that can occur if range made before metal cast. This is two different castings of aluminum.

Chapter 8 Coping with the Unexpected

- Work with reputable firms
- Establish acceptable parameters
- Recognize that anodized surfaces are not coatings
- Establish an acceptable range with the actual metal to be used

WELDING OF ANODIZED ASSEMBLIES

As described, anodizing is not a coating like paint but a conversion of the surface. You can weld aluminum after anodizing but you will have to remove the anodized coating in order to make the weld. Additionally, the heat of welding can fracture the anodizing near the weld area as the metal expands and contracts.

You can anodized after welding, but the color at the weld will be different. Clear anodizing as well as color anodizing will appear differently where the weld has been placed. The design should attempt to reduce the visibility of the weld as much as possible (Figure 8.13).

FIGURE 8.13 Discoloration and porosity around the welds.

SPALLING

In the two-step anodizing process, sometimes a small, circular portion of the outer surface will come off. This is called spalling, and it cannot be repaired. Essentially spalling occurs when hydrogen gas forms and enters the pores of the metal and permeates the oxide layer. This layer delaminates in a round form. As more metal is deposited into the pore the hydrogen migration is deterred and spalling won't occur.

WATER STAINS

Water staining of mill finish aluminum will occur if moisture is allowed to enter between surfaces of stacked sheet or plate. It can also occur on formed parts that are stacked, surface to surface. Stains go from white to gray to black. The darker the stain, the more severe. Water can enter between two surfaces of aluminum if they are stored out of doors or if condensation develops on the metal and

FIGURE 8.14 Examples of small pits that can appear as the plate is polished.

enters between layers by capillary action. The stain is superficial from a mechanical property point of view but aesthetically it can be an issue. Removal may require mechanical abrasion. There are a few chemical methods, but they involve acid treatments. Many automotive stores have aluminum cast wheel cleaners that contain dilute combinations of acids. On small areas these might be effective.

If the surfaces are to be anodized, the etching process can remove mild water stains, but heavy dark stains will appear as streaks in the final work. Mechanically removing the stain by sanding or bead blasting is necessary to prepare for the anodizing process.

INCLUSIONS VISIBLE AFTER POLISHING PLATE

When polishing large plates of aluminum and thick bars of aluminum, inclusions within the metal surface and just under the surface may appear. These are areas where the metal has trapped voids. These are small pit-like regions on the surface: the more you polish, the more these pits appear. Sometimes sections can spall from the surface.

The frustrating part is that this type of inclusion is difficult to see until it manifests during the polishing process (Figure 8.14).

Most all of these issues can be addressed with up front planning and establishing a clear protocol on what is acceptable and possible. The constraints of working with aluminum are all manageable when working with reputable mill suppliers, extruders, foundries, finishing, and fabrication companies. In terms of the era of modern industrial production, aluminum is still young. It will, however, play a major role in the future of civilization. It is the metal designed with the express purpose of furnishing us with the material for our future.

APPENDIX A

Valuable Information and Specifications for Aluminum in Art and Architecture

American Society of Testing Methods

- ASTM E34 Chemical Analysis of Aluminum and Aluminum Alloys
- ASTM E55 Sampling Wrought Nonferrous Metals and Alloys for Determination of Chemical Composition
- ASTM B 209: Standard Specification for Aluminum and Aluminum Alloy Sheet and Plate
- ASTM B 210: Standard Specification for Aluminum and Aluminum Alloy Drawn Seamless Tubes
- ASTM B 211: Standard Specification for Aluminum and Aluminum Alloy Bar, Rod, and Wire
- ASTM B 221: Standard Specification for Aluminum and Aluminum Alloy Extruded Bars, Rods, Wire, Profiles, and Tubes

Aluminum Association of America

- The Aluminum Association: Aluminum Standards and Data 2013
- AA 45 Designation Systems for Aluminum Finishes
- AA 92 Care of Aluminum
- AA TR3 Guidelines for Minimizing Water Staining of Aluminum
- AA TR7 Guidelines for In-Plant Handling of Aluminum Sheet and Plate

APPENDIX B

Tempers on Wrought Sheet, Extrusion, and Plate for Alloys Considered for Use in Art and Architecture

Wrought alloy	Tempers on sheet, extrusion	Tempers on plate
A91100	0, H12, H14, H16, H18	0, H12, H14, H112
A92014	0, T3, T4, T6	0, T451, T651
A92017	—	
A92024	0, T3, T361, T4, T81, T861	0, T351, T361, T851, T861
A93003	0, H12, H14, H16, H18	0,H12,H14,H112
A93004	0, H32, H34, H36, H38	0,H32,H34,H112
A93105	0, H12, H14, H16, H18, H22, H24, H25, H26, H28	—
A94043	—	—
A95005	0, H12, H14, H16, H18, H32, H34, H36, H38	0, H12, H14, H32, H34, H112
A95050	0, H32, H34, H36, H38	0, H1121
A95052	0, H32, H34, H36, H38	0, H32, H34, H112
A95056	—	—
A95083	0, H32, H116, H321	0, H32, H112, H116, H321
A95086	0, H112, H116, H32, H34, H36, H38	0, H112, H116, H32, H34
A95154	0, H32, H34, H36, H38	0, H32, H34, H112
A95754	0, H12, H14, H16, H22	
A96005	T1, T5	—

Wrought alloy	Tempers on sheet, extrusion	Tempers on plate
A96053	—	—
A96060	T5	—
A96061	0, T4, T6	0, T451, T651
A96063	0, T1, T4, T6, T83, T831, T832	—
A96151	F, T6, T652	—
A96463	T1, T5, T6	—
A97075	T6, T73, T76	0, T651, T7351, T7651

APPENDIX C

European Specifications Relevant to Art and Architecture

EUROPEAN SPECIFICATIONS ON ALUMINUM

EN 15530:2008	General	Aluminium and aluminium alloys – Environmental aspects of aluminium products – General guidelines for their inclusion in standards
EN 12258-1:2012		Aluminium and aluminium alloys – Terms and definitions – Part 1: General terms
EN 12258-2:2004		Aluminium and aluminium alloys – Terms and definitions – Part 2: Chemical analysis
EN 12258–3:2003		Aluminium and aluminium alloys – Terms and definitions – Part 3: Scrap
EN 12258-4:2004		Aluminium and aluminium alloys – Terms and definitions – Part 4: Residues of the aluminium industry
EN 515:1993	Wrought Products	Aluminium and aluminium alloys – Wrought products – Temper designations
EN 573-1:2004		Aluminium and aluminium alloys – Chemical composition and form of wrought products – Part 1: Numerical designation system
EN 573-2:1994		Aluminium and aluminium alloys – Chemical composition and form of wrought products – Part 2: Chemical system based designation system
EN 573-3:2013		Aluminium and aluminium alloys – Chemical composition and form of wrought products – Part 3: Chemical composition and form of products

EN 15530:2008	General	**Aluminium and aluminium alloys – Environmental aspects of aluminium products – General guidelines for their inclusion in standards**
EN 573-5:2007		Aluminium and aluminium alloys – Chemical composition and form wrought products – Part 5: Codification of standardized wrought products
EN 485-1:2016		Aluminium and aluminium alloys – Sheet, strip and plate – Part 1: Technical conditions for inspection and delivery
EN 485-2:2016		Aluminium and alumunum alloys – Sheet, strip and plate – Part 2: Mechanical properties
EN 485-3:2003		Aluminium and aluminium alloys – Sheet, strip and plate – Part 3: Tolerances on dimensions and form for hot-rolled products
EN 485-4:1993		Aluminium and aluminium alloys – Sheet, strip and plate – Part 4: Tolerances on shape and dimensions for cold-rolled products
EN 1396:2015		Aluminium and aluminium alloys – Coil coated sheet and strip for general applications – Specifications
EN 546-1:2006		Aluminium and alumunum alloys – Foil – Part 1: Technical conditions for inspections and delivery
EN 546-2:2006		Aluminium and aluminium alloys – Foil – Part 2: Mechanical properties
EN 546-3:2006		Aluminium and aluminium alloys – Foil – Part 3: Tolerances on dimensions
EN 755-1:2016		Aluminium and aluminium alloys – Extruded rod/bar, tube and profiles – Part 1: Technical conditions for inspection and delivery
EN 755-2:2016		Aluminium and aluminium alloys-Extruded rod/bar, tube and profiles –Part 2: Mechanical properties
EN 755-3:2008		Aluminium and aluminium alloys – Extruded rod/bar, tube and profiles – Part 3: Round bars, tolerances on dimensions and form
EN 755-4:2008		Aluminium and aluminium alloys – Extruded rod/bar, tube and profiles – Part 4: Square bars, tolerances on dimensions and form
EN 755-5:2008		Aluminium and aluminium alloys–Extruded rod/bar, tube and profiles – Part 5: Rectangular bars, tolerances on dimensions and form
EN 755-6:2008		Aluminium and aluminium alloys–Extruded rod/bar, tube and profiles – Part 6: Hexagonal bars, tolerances on dimensions and form

Appendix C European Specifications Relevant to Art and Architecture

EN 15530:2008	General	**Aluminium and aluminium alloys – Environmental aspects of aluminium products – General guidelines for their inclusion in standards**
EN 755 7:2016		Aluminium and aluminium alloys-Extruded rod/bar, tube and profiles – Part 7: Seamless tubes, tolerances of dimensions and form
EN 755-8:2016		Aluminium and aluminum alloys-Extruded rod/bar, tube and profiles – Part 8: Porthole tubes, tolerances on dimensions and form
EN 755-9:2016		Aluminium and aluminium alloys – Extruded rod/bar, tube and profiles – Part 9: Profiles, tolerances on dimensions and form
EN 12020-1:2008		Aluminium and aluminium alloys – Extruded precision profiles in alloys EN AW-6060 and EN AW-6063 – Part 1: Technical conditions for inspection and delivery
EN 12020-2:2008		Aluminium and aluminium alloys – Extruded precision profiles in alloys EN AW-6060 and EN W-6063 – Part 2: Tolerances on dimensions and form
EN 13957:2008		Aluminium and aluminium alloys – Extruded round, coiled tube for general applications – Specification
EN 754-1:2016		Aluminium and aluminium alloys – Cold drawn rod/bar and tube – Part 1: Technical conditions for inspection and delivery
EN 754-2		Aluminium and aluminium alloys – Cold drawn rod/bar and tube – Part 2: Mechanical properties
EN 546-1:2006		Aluminium and aluminum alloys – Foil – Part 1: Technical conditions for inspections and delivery
EN 546-2:2006		Aluminium and aluminium alloys – Foil – Part: 2 Mechanical properties
EN 546-3:2006		Aluminium and aluminium alloys – Foil – Part 3: Tolerances on dimensions
EN 1780-1:2002	Cast Products	Aluminium and aluminium alloys – Designation of alloyed aluminium ingots for re-melting, master alloys and castings – Part 3: Writing rules for chemical composition
EN 1780-2:2002		Aluminium and aluminium alloys – Designation of alloyed aluminium ingots for re-melting, master alloys and castings – Part 2: Chemical symbol based designation system
EN 1780-3:2002		Aluminium and aluminium alloys – Designation of alloyed aluminium ingots for re-melting, master alloys and castings – Part 3: Writing rules for chemical composition
CEN/TR 16748:2014		Aluminium and aluminium alloys – Mechanical potential of AI-Si alloys for high pressure, low pressure and gravity die casting

EN 15530:2008	General	**Aluminium and aluminium alloys – Environmental aspects of aluminium products – General guidelines for their inclusion in standards**
CEN/TR 16749:2014		Aluminium and aluminium alloys – Classification of defects and imperfections in high pressure, low pressure and gravity die cast products
EN 1559-4:2015		Founding – technical conditions of delivery – Part 4: Additional requirements for aluminium alloy castings
EN ISO 2085:2010	Anodized Finishes	Anodizing of aluminium and its alloys – Check for continuity of thin anodic oxidation coatings – Copper sulfate test (ISO 2085:2010)
EN ISO 2106:2011		Anodizing of aluminium and its alloys – Determination of mass per unit area (surface density) of anodic oxidation coatings – Gravimetric method (ISO 2106:2011)
EN ISO 2128:2010		Anodizing of aluminium and its alloys – Determination of thickness of anodic oxidation coatings – Non-destructive measurement by split-beam microscope (ISO 2128:2010)
EN ISO 2143:2010		Anodizing of aluminium and its alloys – Estimation of loss of absorptive power of anodic oxidation coatings after sealing – Dye-spot test with prior acid treatment (ISO 2143:2010)
EN ISO 3211:2010		Anodizing of aluminium and its alloys – Assessment of resistance of anodic oxidation coatings to cracking by deformation (ISO 3211:2010)
EN ISO 6581:2010		Anodizing of aluminium and its alloys – Determination of the comparative fastness to ultraviolet light and heat of colored anodic oxidation coatings (ISO 6581:2010)
ENISO 6719:2010		Anodizing of aluminium and its alloys – Measurement of reflectance characteristics of aluminium surfaces using integrating-sphere instruments (ISO 6719:2010)
EN ISO 7599:2010		Anodizing of aluminium and its alloys – General specifications for anodic oxidation coatings on aluminium (ISO 7599:2010)
EN ISO 7668: 2010		Anodizing of aluminium and its alloys – Measurement of specular reflectance and specular gloss of anodic oxidation coating at angels of 20 degrees, 45 degrees, 60 degrees or 85 degrees (ISO 7668:2010)
EN ISO 7759: 2010		Anodizing of aluminium and its alloys – Measurement of reflectance characteristics of aluminium surfaces using a gonio-photometer or an abridged gonio-photometer (ISO/FDIS 7759:2010)
EN ISO 8251: 2011		Anodizing of aluminium and its alloys – Measure of abrasion resistance of anodic oxidation coatings (ISO 8251:2011)

APPENDIX D

Alloy Designations for Wrought Aluminum Alloys Used in Art and Architecture

UNS	DOA designation	ISO designation
A91050	1050A	Al99.5
A91060	1060	Al99.6
A91070	1070A	Al99.7
A91100	1100	Al99.0Cu
A92014	2014	AlCu4SiMg
A92017	2017	AlCu4MgSi
A92024	2024	AlCu4Mg1
A93003	3003	AlMn1Cu
A93004	3004	AlMn1Mg1
A93005	3005	AlMn1Mg0.5
A93103	3103	AlMn1
A93105	3105	AlMn0.5Mg0.5
A94043	4043	AlSi5
A95005	5005	AlMg1
A95050	5050	AlMg1.5
A95052	5052	AlMg2.5
A95056	5056	AlMg5Cr
A95083	5083	AlMg4.5Mn0.7
A95086	5086	AlMg4

Appendix D Alloy Designations for Wrought Aluminum Alloys

UNS	DOA designation	ISO designation
A95154	5154	AlMg3.5
A95754	5754	AlMg3
A96005	6005	AlSiMg
A96060	6060	AlMgSi
A96061	6061	AlMg1SiCu
A96063	6063	AlMg0.7Si
A97075	7075	AlZn5.5MgCu

This is a partial list of wrought aluminum alloys that can be used in art and architecture.

The UNS designation stands for Unified Numbering System and is an alloy numbering convention widely used in North America.

The DOA designation is the Declaration of Accord on the Recommendation for an International Designation System for Wrought Aluminum and Wrought Aluminum Alloys. This designation system was recognized by the United States, Argentina, Australia, Austria, Belgium, Brazil, Denmark, Finland, France, Germany, Italy, Japan, the Netherlands, Norway, South Africa, Spain, Sweden, Switzerland, and the United Kingdom. The European Aluminum Association is also signatory.

ISO stands for the International Organization of Standards. ISO uses an alphanumerical system that begins with the major element system and follows with the major alloying elements required, rounded off to the nearest 0.5%, and the secondary element rounded off to the nearest 0.1.

Further Reading

AAMA (1982). *Care and Handling of Architectural Aluminum from Shop to Site*. 10, AAMA (8 Curtain Wall). AAMA.

Altenpohl, D.G. (1998). *Aluminum: Technology, Applications, and Environment – A Profile of a Modern Metal*. The Aluminum Association, Inc.

The Aluminum Association (2002). *Visual Quality Characteristics of Aluminum Sheet and Plate*, 4e. The Aluminum Association.

The American Welding Society (1972). *Welding Aluminum*, Vol. 6e. The American Welding Society.

Ammen, C.W. (1985). *Casting Aluminum*. TAB Books, Inc.

Birdsall, G.W. (1961). *Aluminum Forming*. Reynolds Metal Company.

Carnegie Museum of Art (2000). *Aluminum by Design*. Carnegie Museum of Art.

Edwards, J. (1955). *The Immortal Woodshed*. Dodd, Mead & Company.

Grist, E. (1997). *Collectible Aluminum*. Collector Books, Division of Schroeder Publishing Co., Inc.

Kaiser Aluminum and Chemical Sales, Inc. (1956). *Casting Kaiser Aluminum*, 1e. Kaiser Aluminum & Chemical Sales, Inc.

Kawai, S. (2002). *Anodizing and Coloring of Aluminum Alloys*. ASM International/Metal Finishing Information Services.

Ockner, P. and Pina, L. (1997). *Art Deco Aluminum*. Schiffer Publishing Ltd.

One Street Components (2014). *Backyard Aluminum Casting*, 1e. One Street Components.

Skrabec, Q.R. (2017). *Aluminum in America*. McFarland & Company, Inc.

Stiles, C. (2010). *Anodized!* Lark Books.

Weidlinger, P. (1956). *Aluminum in Modern Architecture*, vol. II. Reynolds Metals Company.

Welch, B.J. (1999). Aluminum smelting overview: aluminum production paths in the new millennium. *Journal of Metal 51* (5).

Wilquin, H. (2001). *Aluminum Architecture, Construction and Details*. Birkhauser.

Index

Page numbers followed by *f* and *t* refer to figures and tables, respectively.

A

A02080 alloy, 83*t*, 86–88
A02950 alloy, 87
A03190 alloy, 83*t*, 84, 130
A03560 alloy, 83*t*, 84–86, 86*f*, 130
A04430 alloy, 83*t*, 88–89
A05180 alloy, 83*t*, 90, 130
A05200 alloy, 83*t*, 90
A05350 alloy, 83*t*, 91, 130
A07120 alloy, 83*t*, 92, 130
A07130 alloy, 83*t*, 92–93, 130
A12010 alloy, 83*t*, 89
A13190 alloy, 83*t*, 84–85
A13560 alloy, 83*t*, 84
A15350 alloy, 83*t*, 91
A24430 alloy, 83*t*, 89
A25350 alloy, 83*t*, 91
A91xxx series alloys
    characteristics of, 38–39
    components of, 34*t*
    corrosion of, 309
    fabricating with, 231
    forms of, 219
    surface finishes for, 104, 106, 110, 130, 141, 158
    types of, 52–53
    welding, 274–276
A91050 alloy, 122
A91080 alloy, 122
A91090 alloy, 122
A91100 alloy, 51*t*, 52, 53, 105, 263, 303
A92xxx series alloys
    characteristics of, 45
    components of, 34*t*
    corrosion for, 303
    forming of, 257
    surface finishes for, 158
    types of, 53–56
    welding, 274, 275, 279
A92014 alloy, 51*t*, 54
A92017 alloy, 51*t*, 55
A92024 alloy, 21, 51*t*, 55–56, 215, 269
A93xxx series alloys
    characteristics of, 38, 50, 52
    cleaning of, 182, 183
    components of, 34*t*
    fabricating with, 231
    surface finishes for, 141
    types of, 57–59
    welding, 274–276
A93003 alloy
    anodizing quality of, 123
    applications of, 97*f*
    characteristics of, 51*t*, 57–58
    corrosion resistance of, 303
    forming, 263
    surface finishes for, 108*f*
A93004 alloy, 51*t*, 58–59
A93105 alloy, 51*t*, 59, 123
A94xxx series alloys
    characteristics of, 45
    components of, 34*t*
    types of, 59–60
    welding, 274–276
A94034 alloy, 51*t*, 60
A94043 alloy, 139, 276, 303
A95xxx series alloys
    appearance issues for, 329
    characteristics of, 35, 38, 50, 52
    components of, 34*t*
    corrosion of, 289, 291, 300, 306, 309

A95xxx series alloys (continued)
  forming, 257
  forms of, 195
  surface finishes for, 104, 110, 130, 141
  types of, 60–70
  welding, 274–276
A95005 alloy
  anodizing quality of, 122
  architectural applications of, 62f–64f, 218f
  characteristics of, 51t, 61–63
  coloring by anodizing for, 130
  forms of, 63–64, 195
  surface finishes for, 105
A95050 alloy, 51t, 65, 123, 130
A95052 alloy
  anodizing quality of, 123
  art applications of, 67f
  characteristics of, 51t, 65–66
  corrosion resistance of, 303, 306
  forming, 263
  forms of, 195, 219
A95056 alloy, 51t, 66–67
A95083 alloy
  anodizing quality of, 122
  applications of, 21
  characteristics of, 51t, 68–69
  corrosion resistance of, 35, 304, 306
A95086 alloy
  applications of, 21
  characteristics of, 51t, 68–69
  corrosion resistance of, 35, 304, 306
  forming, 263
  forms of, 195
A95154 alloy, 51t, 69–70, 123
A95356 alloy, 276
A95454 alloy, 306
A95456 alloy, 306
A95754 alloy, 35, 51t, 70
A96xxx series alloys
  and appearance issues, 329
  characteristics of, 35, 38, 45, 50, 52
  components of, 34t
  corrosion of, 309

  forms for, 195, 204, 205t, 207–208
  surface finishes for, 104, 110, 130, 141
  types of, 70–77
  welding, 274–276
A96005 alloy, 51t, 71, 110f, 205t
A96053 alloy, 51t, 71–72
A96060 alloy, 51t, 72–73, 205t, 215
A96061 alloy
  anodizing quality of, 123
  applications of, 21, 74f
  characteristics of, 51t, 73
  corrosion resistance of, 304
  fastening of, 269
  forms of, 35, 196, 205t, 214, 215
A96063 alloy
  anodizing quality of, 123
  applications of, 21, 22, 75f, 288f
  characteristics of, 51t, 74–75
  corrosion resistance of, 303, 304
  forms of, 35, 205t, 214, 215
  surface finishes for, 105
A96101 alloy, 214
A96151 alloy, 51t, 76, 123
A96262 alloy, 269
A96351 alloy, 303
A96463 alloy, 51t, 76–77, 205t
A97xxx series alloys
  characteristics of, 45
  components of, 34t
  cutting, 246
  forming of, 257
  surface finishes for, 158
  types of, 77–78
  welding, 274, 275, 279
A97075 alloy, 21, 51t, 77–78, 269
A98xxx series alloys, 34t
AA, see Aluminum Association of America
Abrasion, for paint application, 316
Acids, 186, 313, 352. See also specific types
Adhesives, stiffener, 339–340
Aerospace industry
  A92xxx series alloys in, 54–56
  A97xxx series alloys in, 77–78
  aluminum production for, 17–21

Index

selecting alloys for, 34
surface treatments for, 157
Age hardening (aging), 18, 19, 42, 45, 46
Aircraft wire, 72
Airships, 17, 19, 20
Alcan, 140
Alcanadox, 140
Alclad
   for A91xxx series alloys, 38, 52
   applications of alloys with, 97f, 108f, 202, 203
   corrosion resistance with, 291, 298
   in history of aluminum production, 19
   surface finishes for, 106
Alcoa, *see* Aluminum Company of America
Alcohols, as cleaners, 185
Alkalis, 185–186, 313
Alloys, 33–93
   anodizing quality of, 122–123, 127
   cast, 78–93, 359–360
   characteristics and composition of, 33–35
   choosing, 35
   color differences in anodized surfaces due to, 346
   commercially pure, 38–39
   corrosion resistance of, 285–286
   designation system, 39–93
   initial Mill casting, 35–38
   modern production process for, 18–21
   tempers and strengthening, 42–50
   Unified Numbering System, 40–42
   wrought, *see* Wrought alloys
Alloy coloring, 138–140, 140t
Alloy designation system, 39–93
   for cast alloys, 78–93
   for commercially pure alloys, 38–39
   and tempers/strengthening of alloys, 42–50
   of Unified Numbering System, 40–42
   for wrought alloys, 50–78, 361–362
Alumina
   aluminum oxide and, 113
   from bauxite, 11–12, 14
   in current production process, 25
   natural finish for, 162
"Aluminium," use of term, 29
Aluminum
   "aluminium" vs., 29
   for art applications, 22–25
   characteristics and atomic structure of, 1–7, 8f
   current production process, 25–29
   as design material, 7–10
   environmental and hygienic benefits of, 10–14
   history of, 14–16
   life cycle of, 326–327
   origin of modern production process, 16–22
   other architectural metals vs., 30–31
Aluminum Association of America (AA), 39
   cast alloy designations from, 79, 81
   coating designations from, 124
   on directional and non-directional finishes, 102
   specifications from, 353
   wrought alloy designations from, 40, 41, 49
Aluminum Company of America (Alcoa), 16
   and Alclad, 202
   Class I coating designations from, 124
   integral coloring at, 140
   modern production process at, 20–22
   subsidiaries of, 24
   use of term "aluminum" at, 29
Aluminum Cooking Utensil Co., Inc., 24
Aluminum Industrie Aktien Gesellschaft, 16
Aluminum market, 327, 328f
Aluminum oxide, 3, 4
   in anodizing process, 112–113, 116, 119–121, 120f
   as casting impurity, *see* Dross
   clarity of, 347, 348
   corrosion resistance of, 10, 111, 286–287, 293
   removing, for soldering, 272
   weathering of, 287–288
Alusuisse-Lonza, 16
Alzheimer's disease, 14
AMAG, 9f, 162f
American Metals Market, xi
American National Standards Institute (ANSI), 79
American Society for Testing and Materials (ASTM), 175–176, 175t, 353
American Wire Gauge (AWG), 219
Ammonium bifluoride, 314

Ando, Tadao, 102, 103*f*
Angel Hair finish, 105*f*, 112*t*, 343
Angles, as structural shapes, 215
Annealing, 231
Anode
  area of cathode to, 301
  sacrificial, 289, 297, 313
Anodized finishes, 111–144, 121*f*
  for A95xxx series alloys, 60, 64, 66, 68
  appearance issues for, 343–348
  attributes of, 114
  for castings, 160
  cleaning, 186
  coloring for, 130–144, 346–347
  corrosion resistance of, 299
  crazing, 344, 346–347
  cutting through, 246
  dents and scratches in, 332–333
  design considerations with, 144–147
  European specifications for products with, 360
  expectations of surface with, 161, 167–171
  for extrusions, 159
  graffiti removal from, 333, 334
  history of, 115–116
  managing expectations about, 128–129
  microcracks in, 234–235, 238
  paint coatings applied to, 316
  perforating, 253, 254
  removal of, 142, 144, 343
  for structural shapes, 219
  tonal differences in, 346–348
  variations in, 126
  weathering of, 289, 290*f*, 291
  and welding, 273, 350
Anodized quality (AQ) designation, 35, 64
  alloys with superior and good, 122–123
  and expectations about surface appearance, 128–129
  variables in, 126–127
Anodizing plant, appearance issues and variables in, 344, 345*t*, 347*t*
Anodizing process
  batch, 145–146, 146*t*, 147*t*
  challenges with, 343–344, 345*t*–346*t*

chromic acid, 148
clear, 118*f*, 130, 347, 348
coil, 146–147, 147*f*, 147*t*, 169, 349
defects that appear after, 127–128
dye color, 133–138, 136*f*, 334
hardcoat, 123, 130–132
modern, 20
stains appearing in, 330
steps in, 116–119, 120*f*
sulfuric acid in, 114–116, 118, 124
surface layer resulting from, 119–126
two-stage, *see* Two-stage process
for water stained surfaces, 311, 352
ANSI, *see* American National Standards Institute
Anti-icing salts, 291, 292, 303, 305
Appearance
  issues with, *see* Unexpected appearance conditions
  and life cycle, 326–327
  of matte surfaces, 165
  and texture, 172
AQ designation, *see* Anodized quality designation
Aqua blasting, 102
Architectural metals, 30–31, 30*t*
Areas, ratio of, 300, 301
Art Deco style, 22–24, 23*f*
Artificial aging, 45, 46
Arts Decoratifs et Industriels Modernes exhibition, 22
Asada, Tahei, 115, 141
Asada Process, 141
"As fabricated" finish, 96, 100, 101
Asperities, 309, 322
ASTM, *see* American Society for Testing and Materials
Atomic structure, 1–7
Automotive industry, 76
AWG, *see* American Wire Gauge

B

Back gauge, 239
Backing dies, 211
Bacteria, on aluminum surfaces, 14
Band saws, 240
Bar aluminum, 215

Bartle Hall sculptures (Fischer), 164–165, 164*f*
Bascal, 115
Batch anodizing, 145–146, 146*t*, 147*t*
Bauxite, 3, 10–12
Bed length, 256–257
Bends, microcracks at, 234, 234*f*
Bend radii, 343
Bengough-Stuart process, 139
Beverage cans
   alloys used for, 59
   in aluminum production process, 25, 26
   energy for refining aluminum vs. recycling, 10, 13–14
   quality of aluminum in, xi
Billet
   in casting process, 17
   color differences due to, 346
   for extrusion, 28, 204, 206, 207
   extrusion ratio and cross-section/area of, 210, 211
   foil from, 200
Bimetallic corrosion, *see* Galvanic corrosion
Blanks, spinning, 263*f*
Blimps, *see* Airships
Bloomberg Center (New York, New York), 151*f*
Blue Wool Scale, 175–176, 175*t*
BNIM Architects, 183*f*
Boeing B29 Bomber, 21
Boeing B29 Super Fortress, 78
Bolster dies, 211
"Both sides prime" condition, 235
Bowing, 331*t*
Brazing, 271–272
Bright dipping, 109–111
Brown and Sharpe wire gauge, 219
Buckles, 331*t*, 336
Buffing, 107, 108
Building industry, alloys for, 57–59
Bumping, 223, 224*f*

C

Caboni, P., 133
Capacitor discharge (CD) stud welding, 277, 278*f*
Carbon dioxide, from refinement, 13
Carbon dioxide lasers, 248–249
Carboni, V., 115
Carroll, Lewis, 33
Casey, Thomas Lincoln, 16
Cast alloys, 78–93
   applications for, 82–84, 189, 224–229
   designations for, 78–81
   finishes on, 97, 160
   magnesium, 89–91
   silicon and silicon copper, 84–89
   specifications for, 359–360
   zinc, 92–93
Casting, 280*f*. *See also* Sand casting
   appearance issues from, 340–341
   centrifugal, 225, 226*t*
   die, 90, 226*t*, 228, 229*f*
   double die, 228, 229*f*
   in fabrication process, 279–282
   initial Mill, for alloys, 35–38
   investment, 226*t*
   modern process for, 17
   near net-shape, 78
   permanent mold, 84–86, 89, 92–93, 226*t*, 228
   in production process, 27
Cathode, area of, 301
Cathodic protection, 304, 315
Caustic etching, 117, 169
CCS, *see* Circumscribing circle size
CD stud welding, *see* Capacitor discharge stud welding
Cement, Portland, 313
Centrifugal casting, 225, 226*t*
Channels, 215
Chemical inhibition, 315
Chemical milled surface
   fabricating, 254, 255*f*
   finish of, 112*t*, 157–158
Chemical pretreatments, for painting, 316–317
China, 11–12, 25
Chloride salts, 291, 292, 303
Chromate pretreatments, for painting, 151, 316–317
Chromic acid anodizing, 148
Chromium, 39

The Chrysalis at Merriweather Park (Fornes), 150*f*
Chrysler Building (New York, New York), 17
CIE LAB values, 348
Circular sawing, 240, 242
Circumscribing circle size (CCS), 208, 210, 210*f*
Clad aluminum, *see* Alclad
Clarity, aluminum oxide, 347, 348
Class I anodized coatings, 122, 124, 126
Class II anodized coatings, 124, 125*f*
Clay, 3, 312
Cleaners
    corrosion caused by, 290–291
    pH of, 184–186, 185*f*
    for removing corrosion, 313, 314
Cleaning
    for anodizing, 117
    and corrosion, 290–293, 313–314
    and expectations of surface, 182–186
    of mill finish, 99
    of mirror finish, 108–109
    of natural surfaces, 163, 164, 166
Clear anodizing, 118*f*, 130, 347, 348
Clydes Wine and Dine (New York, New York), 316*f*
CNC, *see* Computer numerical control
Coatings. *See also* Paint coatings
    for cathodes, 301
    Class I, 122, 124, 126
    Class II, 124, 125*f*
    porcelain, 161–162
    on Space Shuttle, 316
Coefficient of thermal expansion, 148, 336, 337*t*
Coherency, 43, 43*f*
Coils, aluminum, 197–200, 198*f*, 311
Coil anodizing
    color matching and, 169, 349
    surface finish of, 146–147, 147*f*, 147*t*
Coil coating
    corrosion resistance with, 317, 317*t*, 320, 321
    surface finish from, 149–150, 151*f*
Coil setting, 331*t*
Cold rolling, 48, 127, 128
Cold sealing, 344

Cold working, 8
    alloys strengthened by, 44, 48, 50*t*
    for coils and sheets, 198
    process of strengthening with, 42–43, 46–50
Color (coloring), 130–144
    alloy coloring, 138–140, 140*t*
    for anodized surfaces, 118–119, 168, 168*f*, 169*f*
    of architectural metals, 30*t*
    and clear anodizing, 130
    and color anodizing, 132–133
    differences in, 346–347
    dye color anodizing, 133–138
    electrolytic coloring, 141–142
    evaluating consistency of, 176–177
    expectations related to, 173–178
    fading of, 174–176, 175*f*, 175*t*, 176*t*
    hardcoat anodizing, 130–132
    integral coloring, 140–141
    and removal of anodized surface, 142, 144
    variations in, 144–145
Color anodized finishes
    alloys for, 64
    dye coloring, 133–138
    process for, 132–133
Colorimeters, 348
Columbia University (New York, New York), 131*f*
Columnar strength, 336, 338
Commercially pure alloys, 38–39, 52–53
Computer numerical control (CNC), 243, 253
Concrete, exposure to, 313
Conductivity
    electrical, 302, 302*t*
    thermal, 273, 274
Contraction to point of fixity, 336, 338, 338*f*
Conversion coating, 315
Copper, xii
    aluminum vs., 1, 5, 30, 30*t*, 31*t*, 33
    appearance of alloys with, 169
    in cast alloys, 84*t*
    and corrosion, 286, 289
    in polluted water, 312
    in wrought alloys, 39, 53–56
Cornell University, 252*f*

Corona gun, 152
Corrosion, 285–323, 314t. *See also* Galvanic corrosion
  of alloys, 19
  and cleaning, 313–314
  and crazing, 346
  crevice, 294t, 310
  exfoliation, 289, 290, 294t, 306–307, 307f
  expectations about, 311–313
  of fasteners, 269, 269f
  filiform, 294t, 307–308, 308f
  fingerprinting, 311
  fretting, 294t, 309–310
  intergranular, 303–306, 305f
  and life cycle of aluminum, 325
  and natural weathering, 287–293
  and paint coatings, 315–321
  pitting, 294t, 303
  protective measures to prevent, 314–315
  removing, 314
  storage and handling to prevent, 322–323
  stress cracking, 294t, 308–309, 309f
  types of, 293–310
  uniform, 294t, 295
  water staining and, 310–311
Corrosion cell, 296f
Corrosion fatigue, 294t
Corrosion resistance, 10, 155, 202, 203
Corundum, 3, 113, 116
Cracking. *See also* Microcracks
  and cutting method, 242
  hot, 276
  spider, 238
  stress corrosion, 294t, 308–309, 309f
Crawford Architects, 218f, 221f
Crazing, 344, 346–347
Creases, 331t
Crevice corrosion, 294t, 310
Cross-section design, extrusion, 208–211
Cryolite, 12, 16
Crystal lattice, 43, 43f, 48

Custom finishes, 101f, 112t
Custom painting, 150–152
Cutoff saws, 240
Cutting, 238–254
  laser, 248–251, 252f
  machining, 246–247
  and microcracks, 332–333, 333f
  plasma, 251
  punching/perforating, 252–254
  routing, 247–248
  saw, 238f, 240–243
  shearing, 239–240
  v-cutting, 242, 242f, 243f
  waterjet, 240f, 243–246, 244f, 245f

D

Dahlquist, David, 318f
Davy, Humphrey, 29
Declaration on Accord on the Recommendation for an International Designation System for Wrought Aluminum and Wrought Aluminum Alloys (DOA) designation, 361–362
Decoiling, sheet, 199, 200f
Deflection, 216, 217f
Degassing, 36
Degreasing, 117
Deionized water, 184–185, 292, 311, 313
Delamination, 333, 351
Delta E values, 349
Dents, 332–333
Deville, Henri Sainte-Claire, 14–15
Dialysis dementia, 14
Dies
  extrusion, 208, 211–213
  marks from, 343
  for press brake forming, 254–260
Die casting, 90, 226t, 228, 229f
Difficulty factor, 211
Direct Chill Process, 17, 20, 21f
Direct extrusion process, 206
Directional surface finishes, 102
Dissimilar metal corrosion, *see* Galvanic corrosion
Distortions, 330–332, 339–340

DOA designation, *see* Declaration on Accord on the Recommendation for an International Designation System for Wrought Aluminum and Wrought Aluminum Alloys designation
Double die casting, 228, 229*f*
Drawn-arc stud welding, 277
Drilling, 148, 249, 250*f*
Drop (heat production step), 36
Dross, 11, 36, 226, 282, 340
Dry molds, 227
Dual curves, 262
Ductility, 10, 231–232
Duh, Chris, 100*f*
Duralumin, 19, 56, 78, 115
Duranodic, 140
Dye color anodizing, 133–138, 136*f*, 334

E
Edge belling, 331*t*
EDM, *see* Electrical discharge machining
Einstein, Albert, 95
Elasticity, modulus of, 232–233, 232*t*, 233*t*
Electrical conductivity, 302, 302*t*
Electrical discharge machining (EDM), 211
Electrolysis, in production process, 16, 29
Electrolytes, strength of, 302–303
Electrolytic coloring, *see* Two-stage process
Electromotive force (EMF), 270*t*, 285, 286*f*, 315n.3
Electromotive scale, 297, 298*t*
Electrons, valence, 3–4
Electropolishing, 111
Electro-potential, 286*t*
Electrostatic painting, *see* Powder coating
Embossed finish, 112*t*, 223, 253*f*
Emerson College, 259*f*
EMF, *see* Electromotive force
Emissivity, 112n.1
Empire State Building (New York, New York), 17, 87
Energy use, with aluminum production, 12
Ennor, William, 17, 20
Environment
  anodic finish and variables in, 169–171, 171*t*
  appearance issues due to, 344, 346*t*
  corrosion protection based on, 314–315
  effects of exposure to, 166*f*
  environmental issues with aluminum use, 10–14
  expectations about effects of, 311–313
ESD, *see* Extra super Duralumin
Etching
  caustic, 117, 169
  chemical, 254
  to remove water staining, 311
  surface finishes featuring, 112*t*, 145*f*, 157, 159*f*
Europe, 41–42, 357–360
Exfoliation corrosion, 289, 290, 294*t*, 306–307, 307*f*
Expanded metal, 220–222
Expansion from point of fixity, 336, 338, 338*f*
Expectations of surface, 161–186
  with anodic finish, 128–129, 167–171
  and cleaning, 182–186
  color-related, 173–178
  and environmental exposure, 311–313
  flatness-related, 178–182
  with natural finish, 162–167
  procedures for improving, 174*t*
  and use of range samples, 171–173
Extra super Duralumin (ESD), 78
Extrusions, 203–213, 204*f*
  anodized surface for, 128
  anodizing, 123
  appearance issues in, 341–343
  challenges in fabricating, 236
  choosing alloys for, 35
  corrosion of, 289, 290
  cross-section design criteria for, 208–211
  designing shape of, 208, 213*f*
  dies for, 211–213
  finishes on, 159–160
  flatness for, 181, 181*f*
  mill finish surface for, 96
  mirror finish for, 108
  modern production process for, 20
  process for making, 207–208, 212*f*
  storage and handling, 322
  supply constraints on, 192–193
  tempers on, 355–356
Extrusion ratio, 210

F

Fabrication facility, quality concerns for, 331–332, 332*t*
Fabrication process, 231–283
  brazing, 271–272
  casting, 279–282
  challenges with, 233–235
  chemical milling, 254
  cutting, 238–254
  fastening, 269–271
  forming, 254–268
  handling and storage, 235–254
  for other architectural metals vs., 31, 31*t*
  rapid prototype process, 283
  soldering, 272
  temper in, 49
  variables in, 169–171, 171*t*
  welding, 272–279
Faraday, Michael, 315n.3
Faraday cage effect, 9, 153, 317
Faraday's laws, 301
Fasteners and fastening, 269–271, 269*f*, 300
Federal Reserve Bank (Boston, Massachusetts), 11*f*
Federal Triangle Flowers (Robin), 82*f*
Feeder plates, 211
Ferdinand, Hereditary Prince of Denmark, 22
Fiber lasers, 249, 250*f*, 252*f*
Filiform corrosion, 294*t*, 307–308, 308*f*
Filler metals, 276
Fingerprints
  and corrosion, 295, 311
  on mill finish surface, 99
  on mirror finishes, 108, 109
  on non-directional finishes, 102–104
Fischer, Ron, 23*f*, 158*f*, 164*f*, 264*f*, 265*f*
Fisheyes, 331, 332*t*
Five-axis milling, 247
Flatness
  criteria for, 335, 337*f*
  expectations about, 178–182
  issues related to, 334–339
Flattened expanded metal, 220
Flaws, in extrusions, 324*f*, 341

Flexibility, of anodized surfaces, 148
Fluidity, 225, 281
Fluorocarbon finish, 320*f*
The Flying Robot (Duh), 100*f*
Foam, aluminum, 189, 229–230, 230*f*
Foil, 200–202
Folded aluminum, 258*f*, 259*f*
Ford Tri-Motor plane, 56, 56*f*
Forging, 76, 268
Formed expanded metal, 220
Forming, 254–268
  choosing alloys for, 35
  forging, 268
  hydro-, 262
  imperfections in, 343
  microcracks due to, 333
  press brake, 221, 222, 254–260, 343
  roll, 260–261
  spinning, 263–264
  stamping, 264–265
  stretch, 261–262
  superplastic, 265–268
Forms of aluminum, 7, 187–230, 241*f*
  basic, 189
  cast, 189, 224–229
  clad, 202–203
  expanded metal, 220–222
  extrusion, 203–213
  foam, 189, 229–230
  foil, 200–202
  and the heat, 190
  other architectural metals vs., 31, 31*t*
  perforated and embossed, 222–224
  pipe and tube, 213–215
  plate, 190–197
  rod and bar, 215
  sheet, 190, 197–200
  structural shapes, 215–219
  and supply constraints, 191–193
  wire, 219
  wrought, 189–224
Fornes, Mark, 150*f*
Foster Partners, 125*f*, 163*f*
Francis of Assisi, St., 231

Fresh water, exposure to, 311–312
Fretting corrosion, 294t, 309–310
Friction stir welding, 279
Frishmuth, William, 15, 16
F temper, 44, 44t
Fuller, Buckminster, 161
Full hard temper, 49
Fusion studs, 277–279, 278f, 340

G
G.A. Dick Corporation, 20
Galling, 127, 235, 322
Galvanic corrosion, 289, 294t, 296–303
   and electrolyte strength, 302–303
   of fasteners, 269, 269f
   oxygen for, 300
   parameters for, 298–300
   preventing, 300f
   ratio of areas for, 300–301
   requirements for, 297f
Galvanic series, 297, 298t
Gasket grooves, in extrusions, 206, 206f
Gemstones, 2, 2f
Gibbsite, 287
Glass bead finish, 101f, 102, 103f, 112t, 343–344
Gloss, 176, 347, 349
Gloss meters, 349
GMAW welding, 273–274
Gold, xii, 1, 15
Gould Evans, 113f
Grace Farms (New Canaan, Connecticut), 62f, 147f
Graffiti removal, 333–334, 336f
Grain, 122, 128, 129, 195f
GTAW welding, 273
Guillotine shears, 239

H
Haas Moto Museum (Dallas, Texas), 63f, 137f
Hadid, Zaha, 156f
Hall, Charles Martin, 16, 29
Hall, Julia Brainerd, 16
Hall-Héroult Process, 16, 22
Hammered finish, 112t

Handling, 235–254, 322–323, 331
Hardcoat anodizing, 123, 130–132
The heat
   appearance issues and, 328
   color variations due to, 145, 346
   in initial mill casting, 35–38
   in production process, 26–27
   for wrought forms, 190
Heat-treatable alloys, 43–45, 47t, 274, 275
Heat treatment, 19, 45
Heller Hostess Ware, 115, 116f
Helmets, 22
Hendrix, Jan, 24f, 67f, 74f, 109f, 110f, 154f, 244f, 321f
Héroult, Paul, 16
Herzog and de Meuron, 220f
Hindenburg zeppelin, 19f
Hollow shapes, extruding, 208, 209f
Holtzman, Malcolm, 157f
Hornbostel, Henry, 17
Hot cracking, 276
Hot rolling, 128
Hot shortening, 341
H temper, 44, 44t, 48–49, 50t
Hydraulic extrusion, 20
Hydrochloric acid, 313
Hydrofluoric acid, 311, 313
Hydroforming, 262
Hydrogen, 273, 340–341
Hydrothermic sealing, 344
Hygienic benefits of aluminum, 10–14

I
I-beams, 215
Image printing, 134f
Inclusions, after polishing, 352. *See also* Pitting corrosion
Industrial applications, anodized coating for, 126
Inertia, moment of, 216, 217f
Ingots
   appearance issues and, 328
   classification of, 80
   and forms of aluminum, 190
   in production process, 27–28, 28f

Inorganic coatings, 161–162
Inorganic dye(s)
   color fading for, 175–176
   coloring with organic dyes vs., 118*f*, 119
   most commonly used, 137n.3
   pigment precipitation with, 134–138
Integral coloring, 140–141
Intergranular corrosion, 303–306, 305*f*
Intermetallic compounds, electro-potential of, 286*t*
International Organization of Standards (ISO) designation, 361–362
Investment casting, 226*t*
Iron alloys
   cast, 84*t*
   characteristics of, 39
   corrosion of, 285, 286, 289
   density of, 5
   tonal differences in, 348
Isabella (Chicago, Illinois), 17
ISO designation, *see* International Organization of Standards designation
IwamotoScott Architecture, 260*f*

J
James Wines Architect, 149*f*
Japan, 115
Joining, of foil, 201

K
Kaiser Aluminum, 20, 22, 140
Kalcolor, 140
Kem Studios, 245*f*
Kensington Company, 24
Kerf, 243, 244
Kyu Sung Woo Architects, 258*f*

L
Laser cutting, 248–251, 252*f*
Laser sintering, 283
Leveling, 180, 181, 200*f*, 261, 331
LeWitt, Sol, 216*f*
Life cycle, product, 326–327
Light absorption, laser cutting and, 249

Light fastness, 175–176, 175*t*
Light Filter (Thomas), 105*f*
Lighting, 20, 221, 222*f*, 349
Light interference, 142, 144*f*
Lime, 313
Liquid spray coating, 317*t*, 320, 321
Lithium alloys, 34
London Airport, 268*f*
Lost wax technique, 225
Lots, ingot, 328
Louvre (Lens, France), 9*f*
Lubricants
   and anodized finishes, 127–128
   flaws in extrusions due to, 341, 342
Lüder lines, 195, 329, 330*f*
Luster, metallic, 173–175

M
Machined aluminum
   fabricating, 246–247, 247*f*, 248*f*
   surface of, 158–159, 160*f*
Magnesium alloys
   cast, 84, 84*t*, 89–91
   corrosion of, 289, 297
   wrought, 39, 60–77
Manganese, 39, 57–59
Marine environments, 10, 35
   A95xxx series alloys for, 68–70
   anodized coating thickness for, 126
   corrosion in, 302–306, 304*f*
   natural weathering in, 288, 289–290
Marks, tool or die, 343
Marring, 235–237, 331*t*, 332, 332*t*
Masking, 145*f*
Matte surfaces, 165, 182
Mechanical cleaning, 186
Medary, Milton B., 326
Medium shrinkage, 225, 281
Melting point, 6*t*, 271–272
Mercury, 312
Mesh, aluminum, 219
Mesh count, 219
Metalabs, 64*f*
Metalix Design, 159*f*
Metallic luster, 173–175

Microcracks, 122, 122f
　in anodized and painted surfaces, 332–333, 333f, 334f
　at bends, 234–235, 234f
　cutting method and, 239
　mishandling and, 238
　welding and, 274
MiG fighter jet, 34
MIG welding, 273–274
Military, aluminum use by, 19–21
Mill
　initial casting by, 35–38
　printing of plate surface by, 196, 196f
　supply constraints at, 191–193, 327–328
　variables in process at, 169–171, 171t
Mill certificate, 37f, 190, 191f, 328
Mill finish surface, 99t
　applications of, 100f
　characteristics of, 96–101
　expectations about, 162–167, 162f, 163f
　graffiti removal from, 333, 334
　other finishes and, 99f, 112t
　stains in, 329, 330
　weathering of, 288f
Milling
　chemical milled surface, 112t, 157–158, 254, 255f
　physical, *see* Machined aluminum
Mill source, variables in, 347t
Mirror finish, 104–109, 109f, 110f, 112t
Mixed Media Artist Proofs (Revenaugh), 135f
Modern Art Museum (Fort Worth, Texas), 102, 103f, 242f
Modulus of elasticity, 232–233, 232t, 233t
Moisture, exposure to, 323. *See also* Water stains
Molds
　dry, 227
　permanent mold casting, 84–86, 89, 92–93, 226t, 228
　rapid prototyping of, 282
　sand casting, 227f, 281f, 282, 282f
Moment of inertia, 216, 217f
Monadnock (Chicago, Illinois), 17
Moneo Vallés, José Rafael, 131f

Morphosis, 151f, 252f, 259f, 316f
Museum Garage (Miami, Florida), 319f
Museum of Science and Industry (Tampa, Florida), 203

N
Nacrite, 293
Narrow shrinkage, 225, 281
NASA, 316
Natural aging, 45, 46
Natural finish, 161–167
Natural weathering, 287–293, 288f
　in marine environment, 289–290
　in Northern urban environment, 291–292
　in rural environment, 292–293
　in Southern urban environment, 290–291
Nd:YAG lasers, 249
Near net-shape casting, 78
Nerman Museum of Contemporary Art (Overland Park, Kansas), 258f
Nickel, 39, 119
Nickel acetate sealing, 344
9/11 Museum (New York, New York), 230, 230f
Nitric acid, 313
No.3 finish, 101
No.8 finish, 101
Nominal dimension, pipe, 213, 214t
Noncoherent structures, 43, 43f
Non-directional surface finishes, 102–104, 167f
Non-heat-treatable alloys, welding of, 274–276
Nonprime side, 98f
Nonspecular surface, 100, 112t
Northern region of United States, weathering in, 291–292
Nybeck, Beth, 104f

O
Oersted, Hans Christian, 14
Off count mesh, 219
Oil, flaws due to, 341, 342
Oil canning, 331t
Olivares, Jonathan, 75f
1-3-5-7-11 sculpture (LeWitt), 216f
Optical profilers, 176
Orange peel, 331t

Ore, xi. *See also* Bauxite
Organic dyes
   color fading for, 174–175
   coloring with, 133–134
   inorganic vs., 118, 118*f*
Organic solvents, 186
O temper, 44, 44*t*
Overbending, 259
Overdyeing, 138*f*, 139*f*, 142, 144*f*
Overtorqueing, 271
Oxygen, aluminum and, 1–3

P
Packaging, 179*f*, 205*f*, 322*f*
Paint coatings, xii, 148–157, 317*t*, 319*f*
   for castings, 160
   for coils, 199
   and corrosion, 307–308, 308*f*, 315–321
   dents/scratches in surfaces with, 333, 334*f*
   expectations of surface with, 161
   for extrusions, 159, 160
   for formed aluminum, 260*f*
   graffiti removal from, 333, 334
   microcracks in, 234–235
   preparing surface for, 314–315
Patinas, 154–157
Peeling, 204
Perfluorocarbon compounds (PFCs), 13
Perforated aluminum
   architectural applications of, 223*f*, 224*f*, 319*f*
   described, 222–223
   fabricating, 252–254, 253*f*, 258*f*
Permanent mold casting
   alloys for, 84–86, 89, 92–93
   other casting methods vs., 226*t*, 228
PFCs, *see* Perfluorocarbon compounds
Phosphate pretreatments for painting, 151
Phosphoric acid, 313
Piercing, microcracks due to, 333
Pigment precipitation, 134–138
Pillowing, 335, 337*f*
Pipe aluminum, 213–214
Pitting corrosion, 303
   after polishing, 351*f*, 352
   in marine environments, 289, 312

   of plate aluminum, 197*f*
   symptoms and remediation for, 294*t*
   in urban environments, 291
Pittsburgh Reduction Company, 16
Plasma cutting, 251
Plate aluminum, 190–197
   appearance issues, 329–332, 330*f*, 334–340
   cutting, 244*f*, 245, 245*f*
   flatness for, 180, 181, 334–339
   mill finish surface for, 97, 98, 100, 101
   mirror finish, 106–108
   roll forming, 262*f*
   supply constraints on, 192, 192*f*
   tempers on, 355–356
   welding, 278*f*
Plating, of aluminum, 161
Platinum, 1
Polished surfaces, cleaning, 186
Polishing
   in anodizing process, 117
   inclusions visible after, 352
   of mirror surfaces, 107, 108
Pollution, 291, 310, 312
Porcelain coatings, 161–162
Porosity, 78–79, 340–341, 350*f*
Portland cement, 313
Potential pH diagram, 294, 295*f*
Pourbaix, Marcel, 293n.1
Pourbaix diagram, 294, 295*f*
Powder coating, 318*f*
   corrosion and, 317
   other paint applications vs., 155*t*, 317*t*, 321
   process for, 152–153
Precipitation hardening, 45, 46
Predock, Antoine, 108*f*
Press brake forming, 221, 222, 254–260, 256*f*, 343
Press quenching, 20
Primary production, 11–13, 25
Primes, 36
Prime side, sheet, 98*f*
Production process
   current, 25–29
   modern, origin of, 16–22
   secondary vs. primary, 10–13

Profiles, 208, 210f, 213. *See also* Extrusions
Profilometers, 176
Punching, 239, 252–254
Pure aluminum
   casting of, 82, 84t
   in wrought alloys, 25, 38

Q
Qatar, 25, 338
Quenching, 20

R
Radial finish, 101f
Range samples, 171–173, 178, 348–350, 349f
Rapid prototyping, 283
RDG Dahlquist Art Studio, 318f
Recycling
   conditions leading to, 327, 327f
   environmental and cost benefits of, 10–11
   recycled material in production process, xi, 11, 25–29
Red mud, 13
Reflectance, 95, 96f, 138
Reflectivity
   of anodized aluminum, 114, 114f
   cutting tool selection based on, 247, 249
   and flatness, 335, 339
   of natural finish, 165–166, 165f, 167f
Reflectors, 8–9, 20, 95, 106
Regazzoni, Ricardo, 106f, 107f
Revenaugh, Katrina, 135f
Reynolds Aluminum, 20, 22
Riley, Terrence, 319f
Riverfront Park Performance Pavilion (North Charleston, South Carolina), 183f
Robert F. Kennedy Department of Justice Building (Washington, DC), 326, 326f, 327
Robin, Stephen, 82f
Rod aluminum, 192, 215
Roll bonding, 203, 203f
Roller leveling, *see* Stretcher leveling
Roll forming, 260–261, 262f
Roping, 331t

Rosenthal Museum of Contemporary Art (Cincinnati, Ohio), 156f
Routing, 247–248
RTV, 316
Rubber pad brake forming, 259–260
Rules of Basketball (Gould Evans), 113f
Rural environment, weathering in, 292–293

S
Sacrificial anode, 289, 297, 313
SAE, *see* Society of Automotive Engineers
SAF, *see* Stabilized aluminum foam
Safety, of aluminum products, 13–14
Salt(s)
   anti-icing, 291, 292, 303, 305
   chloride, 291, 292, 303
   distance from coast and exposure to, 305f
SANAA, 9f, 62f, 147f
Sand casting, 227, 279–282
   alloys for, 84–93
   mold for, 227f, 281f, 282, 282f
   other casting methods vs., 226t, 228
Satin finish, 104, 112t, 167f
Saw cutting, 238f, 240–243
Scalping, 27–28, 190
Scandium, 34
Scrap, 25–26. *See also* Secondary production
Scratches
   in anodized and painted surfaces, 332–333
   in material from original supplier, 331t
   oxide formation after, 315
   preventing, at fabrication facility, 331, 332t
   removing, 163, 236, 237, 343
Screw guide, 271f
Sealing, 119, 344
Seawater exposure, 312. *See also* Marine environments
Secondary production, xi, 11, 25–29
Semi-hollow shapes, 208, 209f
SERA, 224f
Shearing, 239–240, 240f, 333
Shear strength, 270
Sheet aluminum, 190, 197–200
   appearance issues with, 329–332, 334–340
   cutting, 244

flatness for, 180, 181, 334–339
mill finish for, 97–101, 98f, 99f
mirror finish for, 106–108
roll forming of, 260–261
tempers on, 355–356
water staining of, 311
Shielding, for welding, 274
Shipbuilding, 21
Shreve, Lamb, and Harmon, 17
Shrinkage, 225, 226, 280, 281, 341
Shut height, 256
Silage pretreatments, 151
Silicates, 313
Silicon
   aluminum in compounds with, 1–3
   in cast alloys, 82–89, 84t
   in cast alloys with copper, 84–89
   and corrosion resistance, 285
   in wrought alloys, 39, 59–60
Silver, 1, 15, 95
Slats, laser/plasma cutting, 251
Slitting, 239
Smithfield Church (Pittsburgh, Pennsylvania), 17, 18f, 78, 224
Smut, 117
Snohetta, 230, 230f
Society of Automotive Engineers (SAE), 81
Sodium carbonate, 313
Sodium hydroxide, 117, 311
Soils, exposure to, 312
Soldering, 272
Solid shapes, extruding, 208, 209f
Solution heat treatment, 19, 43, 45
Solutions, 45n.1
Solvents, organic, 186
Southern region of United States, weathering in, 290–291
Space Shuttle, 316
Spalling, 351
Specular surface, 100, 106–108, 112t
Spider cracking, 238
Spinning, 263–264, 263f–265f
Springback, 257, 258
Stabilized aluminum foam (SAF), 230

Stains
   mill finish, 329, 330
   water, 237–238, 237f, 310–311, 323, 351–352
Stainless steel
   aluminum alloys vs., 30t, 31t, 33
   appearance of aluminum vs., 166, 167f
   fasteners for aluminum from, 270–271
Stamping, 264–265, 266f, 267f
Standard expanded metal, 220
Steel
   aluminum vs., xii, 30, 30t, 31t
   deflection of, 217f
   embedded particles of, 331–332, 332t
   oxidation of, 293
   plate aluminum vs., 195
   welding, 272, 277
Stiffeners, 181
   and flatness, 335, 337f, 339
   show-through for, 339–340
   welding of, 277
Stocker Hoesterey Montenegro Architects, 63f, 137f
Storage, 235–254, 322–323, 331
Strain hardening, *see* Cold working
Streaks
   and fabrication process, 236, 236f
   structural, 128, 331t, 341, 342f
Stress corrosion cracking, 294t, 308–309, 309f
Stretcher leveling, 180, 181, 331
Stretch forming, 261–262
Strip mining, 11–12
Structural shapes of aluminum, 215–219
Structural streaks, 128, 331t, 341, 342f
The Stubbins Associates, Inc., 11f
Student Center of Qatar University (Doha, Qatar), 154f, 321f
Sulfuric acid, 114–116, 118, 124, 313
Sunburst, 115
Sun Pavilion (Dahlquist), 318f
Superplastic forming, 265–268
Suppliers
   constraints on, 191–193, 327–328
   quality concerns to address with, 330–331, 331t
Support dies, 211

Surface finishes, 6, 95–160, 112*t*. *See also*
    Expectations of surface
  anodizing, 111–144
  bright dipping and electropolishing, 109–111
  for coil and sheet, 199
  design considerations, 144–160
  directional, 102
  for foil, 201, 202*f*
  heat and appearance of, 35, 37–38
  mill finish, 96–101
  mirror, 104–109
  non-directional, 102–104
  for plate, 196, 197*f*
  for structural shapes, 217
Surface marring, 235–237, 331*t*, 332, 332*t*

T
Tampa Museum of Science and Industry (Tampa, Florida), 108*f*
Tange Associates, 229*f*
Tees, 215
Teflon guides, 181*f*
Temper(s), 42–50
  and anodized quality, 123, 128
  for foil, 201
  F temper, 44, 44*t*
  full hard, 49
  heat treatment, 45, 47*t*
  H temper, 44, 44*t*, 48–49, 50*t*
  natural and artificial aging, 46
  O temper, 44, 44*t*
  for plate, 196
  strain hardening, 46–50
  for strengthening, 42–45
  tonal differences due to, 127, 346
  T temper, 44, 44*t*, 46, 47*t*
  on wrought alloys, 355–356
  W temper, 44, 44*t*
Temperature, 8, 227, 282
Texture, 172, 176
Thermal conductivity, 273, 274
Thermal expansion
  coefficient of, 148, 336, 337*t*
  designing for, 338–339
  flatness and, 179–180, 180*f*, 181*f*, 335–339

Thickness
  coil, 198
  cutting tool selection based on, 251, 253, 254
  flatness and, 181–182, 335, 336
  foil, 200
  forming method and, 255
  microcracking and, 333
  plate, 193, 194, 194*t*
  sheet, 198
  supply constraints related to, 192
Thomas, Rachel, 105*f*
3-D printing, 283
TIG welding, 273
Titanium, xii, 30*t*, 31*t*, 39
Tolerances, 169–171
Tone
  anodizing quality and, 127
  differences in, xi–xii, 129, 129*f*, 144–145, 347–348
  expectations about, 169
  and light energy, 172–173, 173*f*
Tool and die marks, 343
Tool steel, 211
T temper, 44, 44*t*, 46, 47*t*
Tube aluminum, 192, 214–215, 275*f*
Turret punches, 253
Two-stage process, 143*f*
  causes of tonal differences in, 347, 348
  color fading for, 175–176
  color produced with, 141–142, 142*t*
  overdyeing and, 138*f*, 139*f*
  spalling in, 351
Type A Lüder lines, 329
Type B Lüder lines, 329

U
Ultraviolet radiation, 174–175
Unexpected appearance conditions, 325–352
  for anodized aluminum, 332–333, 343–348
  from casting process, 340–341
  crazing, 344, 346–347
  for extrusions, 341–343
  flatness-related, 334–339
  graffiti removal, 333–334
  inclusions visible after polishing, 352

Index

and life cycle of aluminum products, 326–327
mill constraints and, 327–328
range samples, 348–350
for sheet and plate aluminum, 329–332, 334–340
spalling, 351
stiffener show-through, 339–340
tolerances and, 169–171
tonal difference, 347–348
water stains, 351–352
welding of anodized assemblies, 350
Unified Numbering System (UNS)
for cast alloys, 79, 81
for wrought alloys, 40–42, 41t, 361–362
Uniform corrosion, 294t, 295
University of Southern Indiana Performing Arts Center (Evansville, Indiana), 157f
UNS, see Unified Numbering System
Urban environments, weathering in, 290–292, 292f
Urethane rubber pad brake forming, 259–260

V

Valence electrons, 3–4
V-cutting, 242, 242f, 243f
Venetian (Chicago, Illinois), 17
Verne, Jules, 1
Vickers Corporation, 19–20
Viewing angle
and appearance of anodized surfaces, 168, 169f
for evaluating color consistency, 177, 177f
and flatness, 335, 339
Visual flatness, 339, 340
Von Schmidt, Chuck, 276f

W

Warping, 254
Washington Monument (Washington, DC), 15–16, 15f, 224
Water
deionized, 184–185, 292, 311, 313
galvanic corrosion due to, 302
Waterjet cutting, 240f, 243–246, 244f, 245f
Water stains
as form of corrosion, 310–311
from improper handling/storage, 237–238, 237f, 323
removal of, 351–352
Wear-Ever Company, 24
Weathering, see Natural weathering
Welding, 272–279, 275f–277f
alloys for, 60, 61f
of anodized components, 148, 350, 350f
batch anodizing after, 146
capacitor discharge stud, 277, 278f
friction stir, 279
fusion stud, 277–279, 278f, 340
of heat-treatable alloys, 275
of non-heat-treatable alloys, 275–276
and surface finish, 164
Wet blasting, 102
Wet coating, 153–154, 153f, 154f, 155t, 321f
Wide range shrinkage, 225, 226, 281
Width
coil, 198, 199, 199t
cutting tool selection based on, 254
foil, 201
plate, 194t
Wilm, Alfred, 18, 42, 46, 56
Winds of Aphrodite (Zhao), 221f
Windspear Opera House (Dallas, Texas), 125f
Wire, 38, 192, 219
Wire feed, 273–274
Wire mesh, 219
Workac, 319f
World War I, 20
World War II, 20, 21
Wright, Orville and Wilbur, 17
WRNS Architects, 132f, 223f
Wrought alloys, 50–78. *See also specific series and alloys*
choosing, 35
classifications of, 33, 34t
designations for, 50–78, 361–362
European specifications for, 357–359
exfoliation corrosion for, 306
strength of, 7–8
tempers on, 355–356

Wrought forms, 189–224
   appearance issues from production of, 344, 345t–346t
   bar, 215
   clad, 202–203
   coil, 190, 197–200
   expanded metal, 220–222
   extrusion, 203–213
   foil, 200–202
   and the heat, 190
   perforated and embossed, 222–224
   pipe and tube, 213–215
   plate, 190–197
   rod, 215
   sheet, 190, 197–200
   structural shapes, 215–219
   and supply constraints, 191–193
   wire, 219
W temper, 44, 44t

X
X-ray fluorescence (XRF) readings, 26, 26f

Y
YAG lasers, 249
Young's modulus, 232–233, 232t, 233t

Z
Zahner, William, 263f
Zeno Paradox of Achilles, 170n.1
Zhao, Suikang, 221f
Zinc, xii
   in alloys, 39
   aluminum vs., 30t, 31t
   in cast alloys, 84t, 92–93
   corrosion of, 289, 297, 313
   fasteners plated with, 271
ZMC-2 blimp, 202